Current Topics in Microbiology and Immunology

221

Editors

Springer
Berlin
Heidelberg
New York
Barcelona
Budapest
Hong Kong
London
Milan
Paris
Santa Clara
Singapore
Tokyo

Genetic Instability and Tumorigenesis

Edited by M.B. Kastan

With 12 Figures

Michael B. Kastan, M.D., Ph.D.
Johns Hopkins Hospital, Ross 345
720 Rutland Avenue
Baltimore, MD 21205
USA

Cover illustration: Diagram of some of the steps involved in modulating genetic changes and cancer development in eukaryotic cells. Cellular stresses can damage cellular targets and contribute to the development of genetic alterations. The DNA can be directly damaged by base-damaging agents or by oxidative stresses breaking the DNA backbone and damage to cellular organelles, such as mitotic spindles, could lead to abnormal chromosome segregation. A wide variety of gene products influence cellular responses to such lesions by controlling signal transduction, DNA repair and/or cell cycle progression. Damaged DNA can then either: 1) lead to cell death; or 2) be repaired; or 3) result in heritable genetic alterations. The appropriate set of genetic alterations in a given cell type can result in a transformed phenotype (i.e. cancer).

Cover design: Design & Production GmbH, Heidelberg

ISSN 0070-217X
ISBN 3-540-61518-0 Springer-Verlag Berlin Heidelberg New York

Library of Congress Catalog Card Number 15-12910
Printed in Germany

Typesetting: Scientific Publishing Services (P) Ltd, Madras

SPIN: 10495388 27/3020/SPS – 5 4 3 2 1 0 – Printed on acid-free paper

List of Contents

List of Contributors

(Their addresses can be found at the beginning of their respective chapters.)

Genetic Instability and Tumorigenesis: Introduction

M.B. KASTAN

Cancer is a disease resulting from alterations of cellular genes which cause phenotypic changes in somatic cells. Usually, when we think about genetic diseases, we think about inheriting one or two abnormal genes from our parents and these gene abnormalities confer the disease phenotype. In contrast, in the majority of cancers, no such inherited gene abnormalities can be identified (which does not mean that they do not exist) and there is no obvious family history suggesting an inherited disease. The vast majority of genes which are altered in the cancer cells are not transmitted through the germ line, but rather become abnormal in somatic cells sometime during the lifetime of the individual. Thus, the critical question which arises is "how do these genetic changes occur in somatic cells?".

Epidemiologic data suggest that exposure to environmental carcinogens contributes to the genesis of at least 80% of all human cancers (DOLL and PETO 1981). Thus, it is natural to suspect that the genetic changes in somatic cells which contribute to the transformed phenotype arise from DNA damage caused by such exposures. Therefore, understanding how cells respond to DNA-damaging agents is likely to be an important component of our understanding of the genesis of human tumors. There are three possible outcomes for a cell following exposure to a DNA-damaging agent: (a) the cell could repair the damage in a timely fashion so that no physiologically significant mutations are passed on to daughter cells; (b) the cell could die (dead cells cannot go on to form tumors) or never replicate again; or (c) a permanent mutation could result by replication of a damaged DNA template or segregation of damaged chromosomes and such mutations would be passed on to daughter cells. Once the appropriate combination of mutations accumulates in a given cell type, a transformed phenotype could result.

Based on these assumptions, the physiologic factors and the gene products which determine which of these outcomes occur following DNA damage would be predicted to be modulators of the tendencies for malignant transformation. Inherited variability in either the metabolism of environmental carcinogens or in these responses to DNA-damaging agents would thus be expected to alter cancer susceptibility. In fact, there are notable exceptions to the statement above that cancer gene abnormalities occur in somatic cells rather than being inherited. There are a handful of inherited syndromes which predispose individuals to develop cancer. The unique feature of cancer as a heritable genetic disease, however, is that these

Johns Hopkins Hospital, Ross 345, 720 Rutland Avenue, Baltimore, MD 21205, USA inherited gene abnormalities do not give the person the disease of cancer, but rather predispose them to develop the disease. Thus, the inherited gene abnormalities do not generate the transformed phenotype, but somehow set the stage for the genetic changes to occur in the somatic cells which lead to the cellular transformation.

There are two ways to envision how this might happen: (a) the inherited gene abnormality is a rate-limiting step in tumor progression and loss of its function creates a situation in which accumulation of other genetic changes could more easily transform cells; or (b) the inherited genetic change affects the rate at which the somatic genetic changes occur. Interestingly, many of the inherited cancer susceptibility syndromes appear to be defects in the cellular responses to DNA damage (Table 1). There are examples of both homozygous recessive disorders (where the patients have multiple clinical problems in addition to the cancer predisposition) and heterozygous germ line defects (where there is no other discernible phenotypic abnormality other than cancer susceptibility) predisposing to cancer.

Examples of recessive disorders with defects in DNA damage responses include xeroderma pigmentosum (defect in excision repair and susceptibility to UV-induced skin cancers), ataxia telangiectasia (defects in a variety of cellular responses to ionizing radiation and a high incidence of lymphoid malignancies), Fanconi's anemia (defect in repair of DNA cross-links and increased susceptibility to myeloid leukemias), and Bloom's syndrome (increased sister chromatid exchanges after DNA damage and susceptibility to leukemias). Examples of dominantly inherited, heterozygous germ line defects which alter cellular responses to DNA damage and increase cancer risk include hereditary non-polyposis colon cancer (with defects in DNA mismatch repair) and Li-Fraumeni syndrome (with defects in p53). Interestingly, for most of these syndromes, the increased cancer susceptibility tends to be for only one tumor type, suggesting some type of specificity for the particular defective process for the given tissue or in the carcinogenic stimulus for that tissue. An exception to this generalization is that Li-Fraumeni patients are at risk for several different tumor types, suggesting that p53 dysfunction is an important rate-limiting step for carcinogenesis in several (but not all) tissue types. The physiologic function of several other cancer susceptibility genes, such as the *BRCA-1* breast cancer susceptibility gene, are not yet clarified. In addition, there are clearly familial risks for certain other tumor types, such as prostate cancer, where the inherited gene defect remains to be identified. It will not be surprising if other cancer sus-

Table 1. Cancer susceptibility syndromes attributable to abnormalities in DNA damage responses

Cancer susceptibility syndrome/inheritance pattern	Defect/disease(s)
Fanconi's anemia/recessive	Cross-link repair/acute myeloid leukemia
Ataxia telangiectasia/recessive	Ionizing radiation responses/lymphomas
Xeroderma pigmentosum/recessive	Excision repair/skin cancers
Bloom's syndrome/recessive	Increased sister chromatid exchanges/leukemias
Li-Fraumeni syndrome/dominant	Germ line p53 mutations/multiple tumors
Hereditary nonpolyposis colorectal cancer/dominant	Mismatch repair defects/colon, ovary, endometrial carcinomas

ceptibility genes similarly play a role in modulating cellular responses to DNA damage.

It is worth repeating that inheritance of these gene defects is not sufficient for generating a transformed phenotype. These inherited physiologic abnormalities primarily appear to increase the likelihood that the cell will accumulate the other genetic changes which generate the transformed phenotype, i.e., they increase genetic instability. This observation further supports the notion discussed above of the critical role of cellular DNA damage responses in human tumorigenesis. The genes which get altered in this transformation process in the somatic cells are likely to be those involved in cell proliferation (oncogenes and tumor suppressor genes), cell death (apoptosis and anti-apoptosis genes), angiogenesis, basement membrane invasion, and metastasis. This is likely to be true whether the tumor arises in an individual with a known cancer susceptibility syndrome or arises sporadically; the concept is that, with many of these susceptibility syndromes, the chance of generating these somatic genetic errors is increased. The tissue specificity for the various genetic susceptibility syndromes is likely to depend on the role that a particular gene product plays in responding to the particular carcinogens that tissue is exposed to (e.g., xeroderma pigmentosum and UV-induced skin cancers) and on which of the various genes which need to be altered in the somatic cells are rate limiting for transformation in that tissue.

We have learned much about the molecular controls of DNA repair processes, cell cycle perturbations after DNA damage, and cell death decisions in recent years. This book tries to capture many of these insights and focuses on the role of genetic instability in tumor development. Included are basic topics, such as the insights gained by studying cell cycle control and genetic instability in genetically manipulable yeast (discussed by Forest Spencer) and the molecular mechanisms involved in gene amplification, a common type of chromosomal alteration seen in human tumor cells (discussed by Thea Tlsty). A general discussion of the role of increased mutation rates in tumors is presented by Keith Cheng and Larry Loeb. The impact of specific defects in DNA repair on carcinogenesis, with a focus on xeroderma pigmentosum, is covered by James Ford and Philip Hanawalt and a more general treatise on chromosomal instability syndromes and tumor development, with more emphasis on ataxia telangiectasia and the recently identified *ATM* gene, is presented by Steven Meyn. Finally, a comprehensive consideration of the various genes which are actually found altered in human tumors, including some emphasis on the role of mismatch repair genes on cancer susceptibility, is covered in the chapter by Kathleen Cho and Lora Hedrick. Though it is hard to completely cover all aspects of this subject, these chapters should be sufficient to provide the reader with a good conceptual base in the field and hopefully a good sense of the current state of the art in our concepts of the molecular controls of human tumorigenesis.

References

Doll R, Peto R (1981) The causes of cancer: quantitative estimates of avoidable risks of cancer in the United States today. J Natl Cancer Inst 66(6): 1191–1308

Genomic Stability and Instability: A Working Paradigm

K.C. Cheng[1] and L.A. Loeb[2]

1 Introduction

Nearly a century of scientific inquiry into the nature of mutation has offered an exciting glimpse of the forces that cause mutation and drive evolution. New concepts make us realize the magnitude of the challenge of duplicating and distributing an accurate copy of the genome each time a cell divides. For each of the 10^{16} somatic cell divisions that occur in a human lifetime, each copy of the genome must contain an intact sequence of some 3 billion nucleotide base pairs, and each daughter cell must receive the correct number of the 46 chromosomes into which the instruction book of life is parceled. The challenging task of accurate DNA replication and distribution must be accomplished while the cell is under constant attack by forces that damage DNA.

[1]Department of Pathology C7804, Jake Gittlen Cancer Research Institute, Adjunct, Department of Biochemistry and Molecular Biology, The Pennsylvania State College of Medicine, Hershey, PA 17033, USA
[2]Department of Pathology SM-30 and Department of Biochemistry, University of Washington Seattle, WA 98195, USA

Genomic instability refers to normal and abnormal tendencies of cells to undergo mutation. Here we attempt to paint a cohesive picture of our understanding of genomic instability in the context of normal and abnormal biological function. To illustrate the importance of this area of research, we open with a brief overview of the events that led to the exciting finding in 1993 that genomic instability mutations are associated with some inherited human diseases and with many human cancers. We consider cellular mechanisms that provide potential sources for genomic instability, pathways involved in maintaining genomic integrity, and the consequences of mutations in these functions. Methods for measuring mutations are briefly compared, first, in the studies of microsatellite instability and then with respect to genomic instability in human cancer. Finally, we briefly consider some potential directions in the research on genomic instability and their possible benefits to our understanding of cancer and genetic diseases.

2 Historical Perspective

Shortly after the rediscovery of Mendel's work on the nature of inheritance near the turn of the century, Theodor Boveri used sea urchin and *Ascaris* embryos to study the importance of chromosomal stability in maintaining cellular function (Boveri 1902). Work in bacteria and their viruses over the ensuing decades led to the discovery of DNA as the repository of genetic information. Basic genetic and biochemical work led to molecular concepts about mechanisms of mutagenesis. Recent interest in genomic instability has been heightened by the remarkable discovery that gross expansion in the number of CGG triplet repeats is diagnostic for the presence of the Fragile X syndrome (Fu et al. 1992). This finding presaged the discovery of multiple repetitive sequence expansion in other human diseases and suggested the possibility that simple repeat mutations might be a marker for genetic instability in human tumors (Cheng and Loeb 1993).

While early work on rare hereditary cancers suggested that loss of function in a single gene could cause cancer (Knudson 1971), work in the 1980s on the more common epithelial neoplasms, particularly colon cancer (Fearon and Vogelstein 1990) and malignant melanoma (Balaban et al. 1986), indicated that multiple mutations are frequently observed in the most common cancers. One mechanism to account for the number of mutations observed in common cancers is for mutations in genomic stability genes to occur early in the course of tumorigenesis. Mutations in genomic stability genes could engender mutations throughout the genome. The high error rates of DNA polymerases first suggested that errors in DNA replication might be a key mechanism to generate the multiple mutations found in human cancers (Loeb et al. 1974). This, together with Nowell's hypothesis that mutations causing genomic instability would lead to accelerated tumor progression (Nowell 1976), led to predictions that humans carrying genomic instability mutations are cancer prone. Individuals with one mutant allele may lack an initial phenotype, but develop cancer at an early age due to loss of the wild-type allele during growth of somatic tissues (CHENG and LOEB 1993; KNUDSON 1985). Until recently, experimental support for the hypothesis that cancer cells may have mutations in genomic stability genes and exhibit a mutator phenotype (LOEB 1991) has been equivocal. The recent demonstration that mutations in genes responsible for mismatch correction cause forms of familial colon cancer and are associated with microsatellite instability (BRONNER et al. 1994; FISHEL et al. 1993; LEACH et al. 1993; PAPADOPOULOS et al. 1994) provides strong evidence for the importance of a mutator phenotype in cancer. It is important to note that decades of groundwork by basic scientists determining the genetic and biochemical mechanisms of mutagenesis in unicellular organisms made it possible to establish the importance of mutator phenotypes in human cancer.

3 Genomic Stability

In the 1950s and 1960s, interest in the question of how rates of mutation are controlled led to the genetic and biochemical dissection of bacteriophage and bacterial mutants that exhibit elevated rates of mutation (COX 1976; DRAKE et al. 1969). The study of these mutator strains and related DNA repair mutants led to a series of important concepts. First, the accuracy of replication by DNA polymerases is inadequate to account for the observed net level of mutation, even after taking into account proofreading mechanisms exhibited by these enzymes. Second, as a corollary it follows that cells must have special mechanisms for the correction of replication errors. Third, cells are exposed to a plethora of environmental chemicals that damage genomic DNA (AMES and GOLD 1991). Moreover, normal metabolic processes such as oxidative metabolism and inflammation also potentially may produce large numbers of alterations in DNA (AMES and GOLD 1991; KLEBANOFF 1988). Fourth, cells have evolved multiple mechanisms for DNA repair (HANAWALT et al. 1979). Thus, multiple cellular biochemical processes can, if altered, increase genomic instability. A comprehensive review of these functions is given elsewhere (CHENG and LOEB 1993). We present here an updated overview, summarized in Fig. 1, which is biased toward newer aspects of genomic instability.

3.1 Endogenous Sources of Genomic Instability

Genomic stability can be viewed as two opposing sets of reactions that are maintained at equilibrium: one set of reactions alters DNA and the other restores the original sequence. When DNA damage exceeds repair, mutations are manifested. This basic paradigm is embellished by more complex phenomena, such as the induction of DNA damage pathways and the consequences of error-prone DNA repair processes. It provides a useful framework for evaluating the potential of

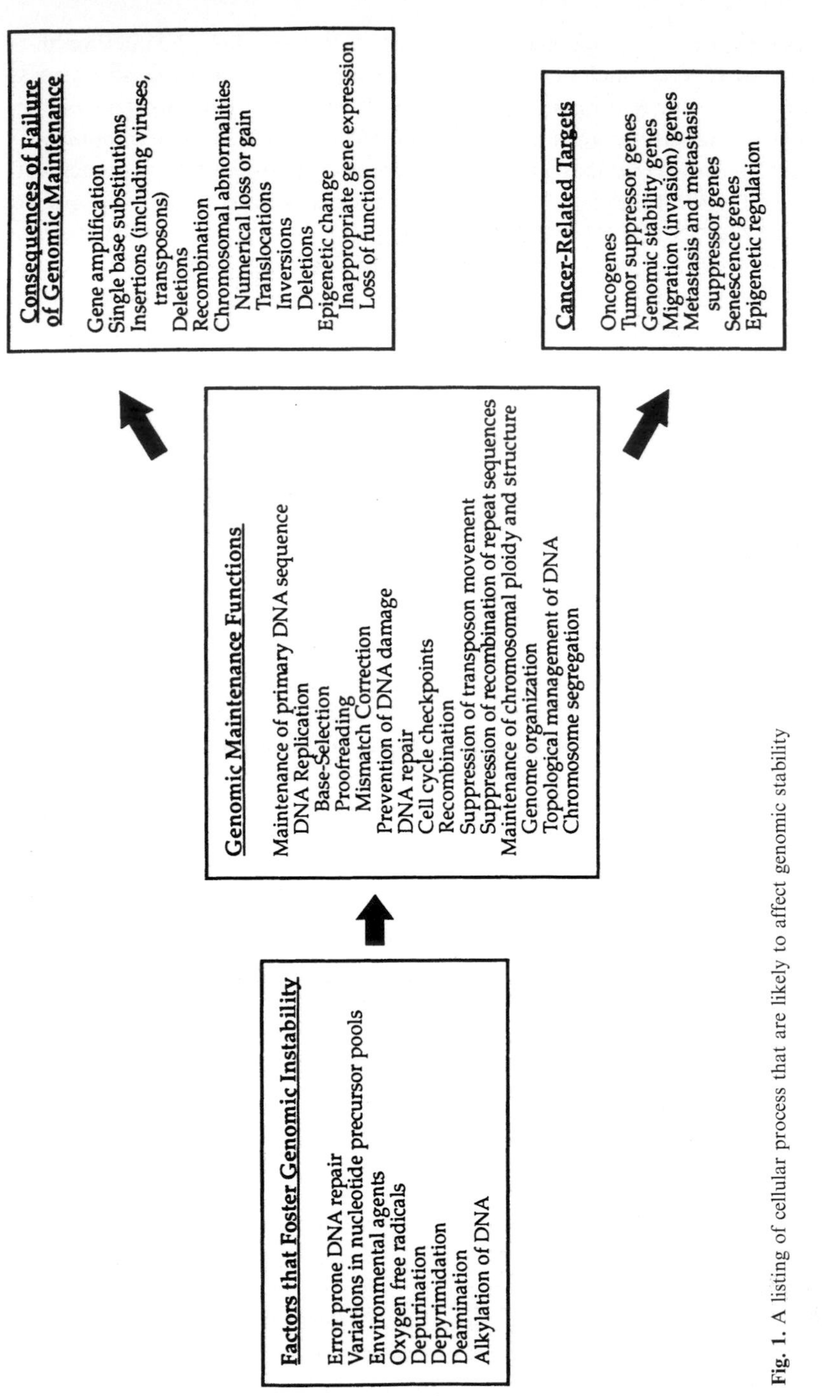

Fig. 1. A listing of cellular process that are likely to affect genomic stability

DNA-damaging events to cause mutations. It has long been recognized that many environmental agents, including irradiation, chemicals, and heat, have the potential to damage DNA in vivo. In somatic cells, these mutations are associated with cancer. Recently, there has been increasing evidence that normal biological processes can also damage DNA or alter the nucleotide sequence in DNA (LOEB and CHENG 1990). If the rates of DNA damage by these exogenous and endogenous sources of DNA damage exceed the corresponding rates of repair, they could contribute to the large number of mutations observed in human cancers. Three normal cellular processes with the potential to cause DNA damage and/or mutations at exceptionally high frequencies will be considered.

3.1.1 Chemical Instability of DNA

In cells, the nucleotide bases in DNA are subject to attack by water. Depurination, the cleavage of the carbon bond between the purine base and its sugar, is the most frequent hydrolytic modification of DNA. It results in the formation of an abasic site which, if not repaired, codes for the incorporation of adenine opposite the abasic site (LOEB and PRESTON 1986). Other hydrolytic reactions, such as depyrimidation and deamination, are less frequent. Based on the rates of depurination of DNA in vitro, it has been estimated that each human cell undergoes 10000 depurinations per day (LINDAHL and NYBERG 1972). Despite the cell's multiple enzymatic activities for the repair of abasic sites, it seems reasonable that some of these abasic sites escape repair and are mutagenic at the time of DNA replication. In spite of the high frequency of depurination, a survey of nucleotide changes that have been reported in genes associated with cancer does not indicate that transversions are the most frequent mutations, suggesting other sources of mutations are more important.

3.1.2 Oxygen Free Radicals

Respiration and a variety of metabolic processes generate as intermediates a series of reactive oxygen species (FLOYD 1990; KLEBANOFF 1988). Based on the excretion of oxygen free radical-modified nucleotides in human urine, it has been estimated that each cell removes from its DNA approximately 20000 oxygen radical-induced modifications per day (CATHCART et al. 1984). The most frequent mutations resulting from oxygen free radical damage to DNA in both prokaryotes and eukaryotes are C to T transitions (FEIG and LOEB 1993). However, the same substitutions are produced by other processes and thus are not pathognomonic of mutagenesis by oxygen free radicals. A much less frequent mutation, a double tandem CC to TT substitution, has so far been reported to result only from damage to DNA by UV irradiation (BRASH et al. 1991) and oxygen free radicals (REID and LOEB 1993). The presence of CC to TT mutations in internal human tumors that are not exposed to UV irradiation would constitute strong evidence for the involvement of oxygen free radicals in pathological processes. So far, this evidence is lacking.

3.1.3 Misincorporation by DNA Polymerases

The frequency of incorporation of non-complementary nucleotides by DNA polymerases is much greater than the observed frequency of mutations found in DNA after replication in cells. Other mechanisms in cells can increase the fidelity of DNA polymerases. These include proteins that interact with DNA polymerases and increase base selection and proofreading (Fry and Loeb 1986) as well as mismatch correction, a mechanism to excise misincorporations left after DNA synthesis (Modrich 1987). A comparison of the error rates of replicating DNA polymerases and the spontaneous mutation rate in normal cells suggests that misincorporation by DNA polymerases in the absence of mismatch correction could account for as many as 6000 mutations per cell division cycle (Loeb 1989). Among the eukaryotic DNA polymerases that have so far been investigated, DNA polymerase-β is by far the most error prone (Kunkel 1985, 1992). Since DNA polymerase-β is involved in DNA repair synthesis, these results suggest that DNA repair itself may be a further source of mutations. Studies on the spectrum of mutations produced by DNA polymerases have not as yet revealed specific mutations that are diagnostic of errors in DNA replication, with the possible exception of slippage during the copying of repetitive nucleotide sequences (Schachman et al. 1960) (see below).

3.2 Deficits in DNA Repair

Considering the plasticity of the DNA molecule and the frequency of DNA damage by both environmental chemicals and endogenous processes, it is not surprising that cells have evolved multiple and redundant pathways for DNA repair. It has been recognized for more than 50 years that there are rare inherited human diseases that manifest both deficits in DNA repair and an increase in the incidence of specific human tumors (Friedberg 1985). Diseases such as xeroderma pigmentosum, Fanconi's anemia, and ataxia telangiectasia are characterized by repair deficiencies in DNA damage induced by specific agents. As a result, exposure of affected individuals to these same DNA-damaging agents is associated with an increased incidence of cancer. What was not adequately appreciated is that deficits in DNA repair might also be associated with the more common human cancers.

3.3 Cell Cycle Checkpoints

An important component in the maintenance of the genome is the coordination and control of DNA replication, repair, and the distribution of DNA to daughter cells during each division cycle. For example, replication of altered DNA prior to removal of modified bases by DNA repair enzymes would lead to miscoding during DNA replication. As in the study of mismatch correction, much of our knowledge about checkpoints in the cell cycle has been gained from an analysis of mutants. Mutants in these processes abolish the normal arrest in the cell cycle that occurs in response to DNA damage. At least two stages of the cell cycle are regulated in response to DNA damage: the G_1-S and G_2-M boundaries (HARTWELL and KASTAN 1994; KASTAN et al. 1992). Cells delay cell cycle progression in order to facilitate the repair of DNA damage and to ensure that previous steps in the cell cycle are complete before proceeding. The relevance of these initial observations in yeast is underscored by the high frequency of abnormalities in p53-related checkpoint functions in human cancer (HARTWELL and KASTAN 1994; KASTAN et al. 1992). Individuals carrying germ line mutations in p53 have a proclivity toward the development of certain tumors (HARTWELL and KASTAN 1994; KASTAN et al. 1992; MALKIN et al. 1990; SRIVASTAVA et al. 1990) and the majority of human cancers have mutations in p53. In addition, mice without p53 have an increased incidence of multiple cancers (DONEHOWER et al. 1992). Perhaps the most incriminating association between p53 mutations and cell cycle delay is that tumor cells lacking p53 do not exhibit the G_1-S delay after exposure to ionizing irradiation, and that the introduction of the wild-type allele of p53 restores the G_1-S delay (KASTAN et al. 1992). The biochemical mechanisms underlying the cell cycle checkpoints for DNA repair are not known, and the coordination of these processes with DNA repair and transcription remains to be determined. Conceivably, there is a complex machinery at the G_1-S phase transition that coordinates cell cycle delays, transcription-coupled DNA repair, and DNA replication.

3.4 Recombination

While recombination has been studied in great detail, primarily in unicellular systems (LOW 1988), there is little known about the enzymological mechanisms for recombination in human cells. Recombinatorial mechanisms have been postulated to play a role in DNA repair, DNA replication, chromosome segregation, evolution, and disease (RADMAN et al. 1993). A major concern with respect to genetic stability is how the cell prevents recombination amidst the enormous repertoire of repetitive DNA elements. For example, there are hundreds of thousands of *alu* repeats in the human genome. If recombination were to occur frequently between these repeats, a myriad of chromosomal transversions, deletions, and inversions would occur. This suggests the hypothesis that specific control mechanisms limit recombination between repeats and that deficiencies in these mechanisms would cause genetic instability.

4 Multiple Mutations and Cancer

Human cancers exhibit multiple chromosomal abnormalities (MITELMAN 1991). Since these chromosomal alterations have been categorized cytologically, they involve large regions of chromosomes that contain millions of nucleotides. Chromosomal aberrations in tumors could be the tip of an iceberg, with a much greater number of smaller mutations, such as single-base substitutions, lying undetected in the sea of DNA sequences (Cheng and Loeb 1993; Loeb 1991); if so, these will be identified increasingly as more sensitive molecular techniques become available. Unfortunately, this hypothesis cannot be immediately tested since we still lack practical methods to measure random single-base substitutions in DNA. A variety of studies suggest that the rate of spontaneous mutagenesis in normal human cells is about 1.4×10^{-10} mutations/nucleotide/cell generation. We calculated that this rate is only sufficient to generate three mutations per cell during a normal lifespan (Loeb 1991). A much higher mutation rate is necessary to account for the six to nine mutations that have been reported in human cancers (Sugimura 1992; Vogelstein et al. 1988) or the much larger numbers of smaller mutations that may be present.

Different mechanisms could account for the multiple mutations that have been found in human tumors. One such mechanism is that an early event in the carcinogenic process causes a mutation in a gene that normally functions in maintaining the stability of the human genome. This mutation then engenders additional mutations throughout the genome. The subsequent random mutations might include some in other genes also involved in maintaining genetic stability. Examples of these early mutations include genes involved in DNA repair, DNA replication, and chromosomal segregation. Mutations involved in oncogenes that are involved in the cancer phenotype might occur later during the course of tumor progression.

Until recently, there has been little evidence to support the concept of a mutator phenotype in cancer. Early efforts to detect mutation relied upon single gene targets, such as *HPRT*. These lack the sensitivity of assays using microsatellites that examine multiple loci and are hypersensitive to mutation. The demonstrations that repetitive nucleotide sequences (microsatellites) are mutated in many different types of human tumors and not in normal cells now provides the strongest evidence for a mutator phenotype (Loeb 1994). Expansion or contraction of repetitive sequences provides an exceptionally sensitive indicator of mutagenesis. If the finding that cancer cells exhibiting microsatellite variation can be generalized to high mutation rates in expressed genes, then microsatellite instability provides a diagnostic marker for a general mutator phenotype in cancer. It should be noted that the mutation rate at the *HPRT* locus has been reported to be 200–600-fold higher in three colon cancer cell lines exhibiting microsatellite instability (Bhattacharyya et al. 1994). The mutation spectra differed, one cell line exhibited 25% frame-shift mutations at a mutational hotspot in the *HPRT* gene, and the other two lacked this hotspot, suggesting that there are at least two mutational pathways leading to microsatellite instability. Thus, microsatellite instability may indeed serve as an excellent diagnostic marker for general mutator phenotypes in cancer.

4.1 Microsatellite Instability and Mismatch Repair

Microsatellite instability was initially observed in tumors from patients with hereditary nonpolyposis colorectal cancer (HNPCC) (AALTONEN et al. 1993). Inheritance in 60% of these families maps to chromosome 2p (AALTONEN et al. 1993; PELTOMAKI et al. 1993), and the protein encoded by this locus is a homolog of the bacterial mutS protein, a key component of the mismatch repair pathway. Based on this homology, it was named hMSH2 (human mutS homolog) (FISHEL et al. 1993). In addition, 30% of patients with hereditary nonpolyposis families segregate with a locus on chromosome 3p (LINDBLOM et al. 1993), encoding a different mismatch repair protein, hMLH1, homologous to the bacterial mutL protein (PAPADOPOULOS et al. 1994). For the basic scientist, the finding that mutations in human homologs of these *Escherichia coli* and yeast mismatch repair genes are associated with human cancer is particularly gratifying. It vividly demonstrates that extensive studies in prokaryotes on fundamental problems can be instrumental in understanding human cancer. Since mismatch repair corrects errors in DNA synthesis in prokaryotes, it has been generally assumed that microsatellite instability is the result of errors by DNA polymerase that are too numerous to be corrected by mutated mismatch repair proteins. It is assumed that the localization of these mutations to repeats could be a manifestation of slippage of DNA during copying of the repetitive nucleotide sequences.

4.2 Microsatellite Instability and Human Cancer

In addition to hereditary nonpolyposis coli, microsatellite instability has been reported with varying frequencies in a number of sporadic tumors. In most of these tumors, it remains to be determined if this instability is also associated with mutations in mismatch repair genes. Human tumors exhibiting microsatellite instability include cancers of the colon (THIBODEAU et al. 1993), stomach (RHYU et al. 1994), and endometrium (RISINGER et al. 1993), and there is evidence for the presence of microsatellite instability in lung cancer (SHRIDHAR et al. 1994). A lack of microsatellite instability has been reported in breast cancers and in male germinal cell cancers (LOTHE et al. 1993). Interestingly, we are unaware of many reports on microsatellite instability in experimentally induced tumors in animals, but we see no reason why these will not be soon forthcoming. Studies of microsatellite instability are reported as being either positive or negative. False positives can be the result of slippage or recombination of truncated products during the polymerase chain reaction (PCR). Also, false positives can be the result of PCR-associated laddering of bands differing by two nucleotide repeat unit steps during amplification of (CA)n microsatellites. As a result, the experiments with high validity are only those that directly and simultaneously compare tumor DNA and normal DNA from the same individual. False negatives, i.e., the lack of expansion, could simply reflect the limited number of microsatellite sequences examined. Microsatellite instability might be evident in most cancers if a sufficiently large number of the 50 000 repetitive sequences in the human genome were probed.

Tumors that exhibit microsatellite instability are frequently designated as RER^+ (Liu et al. 1995), indicating that the mutations are generated by replication errors; however, the source of mutations has not been unequivocally established. For example, heteroduplex DNA yielding nonparental microsatellite alleles could arise not only from errors in DNA synthesis, but also from recombination intermediates. Microsatellite instability may therefore be imagined to be caused by mutations in DNA polymerases or recombination proteins. Moreover, microsatellite instability could be a sensitive indicator for an elevation in mutation frequency in tumors caused by mutations in a variety of other DNA synthetic processes including DNA repair. Complementation assays have already indicated the presence of at least four different mutations that generate microsatellite instability in vitro utilizing extracts from endometrial and colorectal cell lines (Umar et al. 1994). Liu et al. (1995), reported that the mutation in nine of ten cases of sporadic colon cancers which exhibited microsatellite instability was not one of those reported to be associated with mismatch repair. Thus, there are mutations other than the four which have so far been identified in mismatch repair genes that cause microsatellite instability.

The role of genomic instability in cancer may be illustrated by plotting somatic mutation rates over time in spontaneous and hereditary cancers. In Fig. 2, the gray area designates the normal range of mutation. The lower non-zero baseline level is due to the balance between mutational influences (Fig. 1) and repair mechanisms. The range of mutation rate is due to normal allelic and environmental variations. In spontaneous cancer, mutations causing genomic instability may be dominant or recessive. For dominant mutations, the time at which the mutation occurs would coincide with a stepwise increase in mutation rate. For the more common recessive

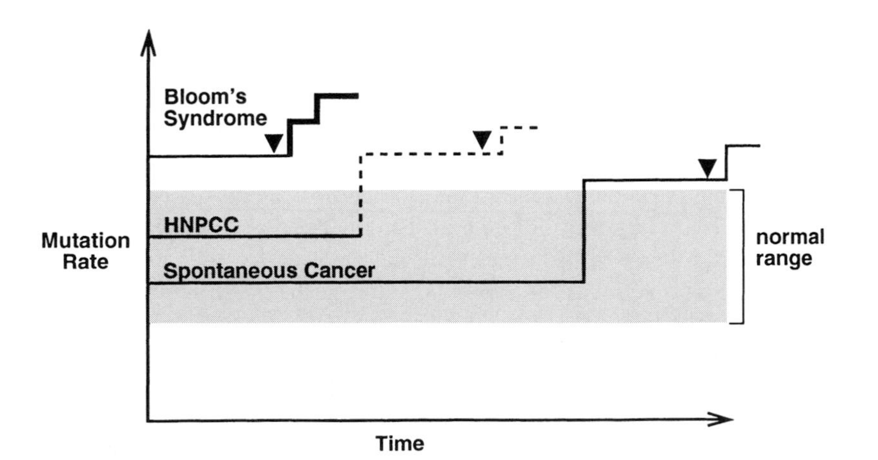

Fig. 2. Postulate increase in mutation rate over time in spontaneous and hereditary human cancers. The *triangle* indicates the clinical manifestation of a malignancy

mutations, two hits are required to elevate the mutation rate. The first mutation arises at an undesignated earlier time, and loss of the remaining wild-type allele is coincident with an increased mutation rate. Loss of expression of the remaining wild-type allele may occur by chromosome loss, recombination, or mutation. Subsequent increases in mutation rate are due to mutations in other genomic stability genes. In inherited genomic instability syndromes, such as Bloom's syndrome (WARREN et al. 1981), the baseline mutation rate is elevated from birth. Secondary mutations in genomic stability genes may or may not be necessary for the development of cancer. In the case of HNPCC, a recessive genomic instability mutation is inherited, and the baseline mutation rate at birth may be normal. Since one copy of a genomic stability gene is already mutated, loss of the wild-type allele occurs earlier, leading to an elevated mutation rate. Alternatively, the extra burden of repair associated with abnormal cellular proliferation (e.g., benign tumor) may cause otherwise recessive genomic instability mutations to become dominant (LIU et al. 1995). The mismatch repair pathway defective in HNPCC might be expected to be a general repair system, and yet there is tissue specificity for tumor susceptibility. The existence of different repair pathways for different tissues may explain this tissue specificity. Relative rates of mutation and types of mutation caused by genomic instability mutations are expected to be determined by the pathway affected.

5 Experimental Systems and Future Directions

The genetic and biochemical analysis of unicellular organisms laid the foundation for the discovery that mutations in mismatch repair genes are associated with human cancer. Further studies in genetically tractable prokaryotes and simple eukaryotes may provide insights into other mechanisms of genomic instability. For example, the pathogenic yeast *Candida albicans* spontaneously switches between alternative phenotypes (SOLL et al. 1953) and African trypanosomes undergo recombination-based antigenic variation (VAN DE PLOEG et al. 1992). Established eukaryotic systems including biochemical approaches with cell extracts and knockout mutations in mice to test homologs for genomic instability phenotypes are expected to continue to yield further insights. Newer vertebrate systems for generating mutants, such as the zebrafish, also may yield more candidate genes to test in human cancers.

New assays which detect other types of mutations or increase sensitivity of mutation detection might reveal that genomic instability is more common than we might expect, and they may also open windows of opportunity to better understand and perhaps interfere with genomic instability in the hope of arresting tumor progression. Thus, the study of genomic instability is likely to provide not only the basic insights into mechanisms of mutation and evolution, but also knowledge relevant to cancer prognosis, drug design, and differential drug sensitivity.

Acknowledgements. This work was supported by grants from the Jake Gittlen Memorial Golf Tournament, the American Cancer Society (IRG-196 and JFRA-581), the Four Diamonds Fund, the National Science Foundation (MCB9319817), and the Central Pennsylvania Oncology Group to KCC, and an Outstanding Investigator Grant from National Cancer Institute, CA39903, to LAL.

References

Aaltonen LA, Peltomaki P, Leach FS, Sistonen P, Pylkkanen L, Mecklin J-P, Jarvinen H, Powell SM, Jin J, Hamilton SR, Petersen GM, Kinzler KW, Vogelstein B, de la Chapelle A (1993) Clues to the pathogenesis of familial colorectal cancer. Science 260: 812–816

Ames BN, Gold LS (1991) Endogenous mutagens and the causes of aging and cancer. Mutat Res 250: 3–16

Balaban GB, Herlyn M, Clark WH Jr, Nowell PC (1986) Karyotypic evolution in human malignant melanoma. Cancer Genet Cytogenet 19: 113–122

Bhattacharyya NP, Skandalis A, Ganesh A, Groden J, Meuth M (1994) Mutator phenotypes in human colorectal carcinoma cell lines. Proc Natl Acad Sci USA 91: 6319–6323

Bohr VA, Smith CA, Okumoto DS, Hanawalt PC (1985) DNA repair in an active gene: removal of pyrimidine dimers from the DHFR gene of CHO cells is much more efficient than in the genome overall. Cell 40: 359–369

Boveri T (1902) Über mehrpolige Mitosen als Mittel zur Analyse des Zellkerns. Verhandlungen der Deutschen Zoologischen Gesellschaft, Würzburg

Brash DE, Rudolph DA, Simon JA, Lin A, McKenna GJ, Baden HP, Halperin AJ, Ponten J (1991) A role for sunlight in skin cancer: UV-induced p53 mutations in squamous cell carcinoma. Proc Natl Acad Sci USA 88: 10124–10128

Bronner CE, Baker SM, Morrison PT, Warren G, Smith LG, Lescoe MK, Kane M, Earabino C, Lipford J, Lindblom A, Tannergard P, Bollag RJ, Godwin AR, Ward DC, Nordenskjold M, Fishel R, Kolodner R, Liskay RM (1994) Mutation in the DNA mismatch repair homologue hMLH1 is associated with hereditary nonpolyposis colon cancer. Nature 368: 258–261

Cathcart R, Scheiers E, Saul RL, Ames BN (1984) Thymine glycol and thymidine glycol in human and rat urine: a possible assay for oxidative DNA damage. Proc Natl Acad Sci USA 81: 5633–5637

Cheng KC, Loeb LA (1993) Genomic instability and tumor progression: mechanistic considerations. Adv Cancer Res 60: 121–156

Cox EC (1976) Bacterial mutator genes and the control of spontaneous mutation. Annu Rev Genet 10: 135–156

Donehower LA, Harvey M, Slagle BL, McArthur MJ, Montgomery CA Jr, Butel JS, Bradley A (1992) Mice deficient for p53 are developmentally normal but susceptible to spontaneous tumors. Nature 356: 215–221

Drake JW, Allen-Forsberg SA, Preparata R-M, Greening EO (1969) Genetic control of mutation rates in bacteriophage T4. Nature 221: 1228–1132

Fearon ER, Vogelstein B (1990) A genetic model for colorectal tumorigenesis. Cell 61: 759–767

Feig DI, Loeb LA (1993) Mechanisms of mutation by oxidative DNA damage: reduced fidelity of mammalian DNA polymerase β. Biochemistry 32: 4466–4473

Fishel R, Lescoe MK, Rao MRS, Copeland NG, Jenkins NA, Garber J, Kane M, Kolodner R (1993) The human mutator gene homolog MSH2 and its association with hereditary nonpolyposis colon cancer. Cell 75: 1027–1038

Floyd RA (1990) Role of oxygen free radicals in carcinogenesis and brain ischemia. FASEB J 4: 2587–2597

Friedberg EC (1985) DNA repair. Freeman, New York

Fry M, Loeb LA (1986) Animal cell DNA polymerases. CRC Press, Boca Raton

Fu Y-H, Pizzuti A, Fenwick RG, King J Jr, Rajnarayan S, Dunne PW, Dubel J, Nasser GA, Ashizawa T, DeJong P, Wieringa B, Korneluk R, Perryman MB, Epstein HF, Caskey CT (1992) An unstable triplet repeat in a gene related to myotonic muscular dystrophy. Science 255: 1256–1259

Hanawalt PC, Cooper PK, Ganesan AK, Smith CA (1979) DNA repair in bacteria and mammalian cells. Annu Rev Biochem 48: 783–836

Hartwell LH, Kastan MB (1994) Cell cycle control and cancer. Science 266: 1821–1827
Kastan MB, Zhan Q, El-Deiry WS, Carrier F, Jacks T, Walsh WV, Plunkett BS, Vogelstein B, Fornace AJ Jr (1992) A mammalian cell cycle checkpoint pathway utilizing p53 and GADD45 is defective in ataxia telangiectasia. Cell 71: 587–597
Klebanoff SJ (1988) Phagocytic cells: products of oxygen metabolism. In: Gallin JI et al (eds) Inflammation: basic principles and clinical correlates. Raven, New York, pp 391–444
Knudson AG Jr (1971) Mutation and cancer: statistical study of retinoblastoma. Proc Natl Acad Sci USA 68: 820–823
Knudson AG Jr (1985) Hereditary cancer, oncogenes and antioncogenes Cancer Res 45: 1437–1443
Kunkel TA (1985) The mutational specificity of DNA polymerase-β during in vitro DNA synthesis. J Biol Chem 280: 5787–5796
Kunkel TA (1992) DNA replication fidelity. J Biol Chem 267: 18251–18254
Leach FS, Nicolaides NC, Papadopoulos N, Liu B, Jen J, Parsons R, Peltomaki R, Sistonen P, Aaltonen LA, Nystrom-Lahti M, Guan X-Y, Zhang J, Meltzer PS, Yu J-W, Kao F-T, Chen DJ, Cerosaletti KM, Fournier REK, Todd S, Lewis T, Leach RJ, Naylor SL, Weissenbach J, Mecklin J-P, Jarvinen H, Petersen GM, Hamilton SR, Green J, Jass J, Watson P, Lynch HT, Trent JM, de la Chapelle A, Kinzler KW, Vogelstein B (1993) Mutations of a mutS homolog in hereditary nonpolyposis colorectal cancer. Cell 75: 1215–1225
Lindahl T, Nyberg B (1972) Rate of depurination of native deoxyribonucleic acid. Biochemistry 11: 3610–3618
Lindblom A, Tannergard P, Werelius B, Nordenskjold M (1993) Genetic mapping of a second locus predisposing to hereditary nonpolyposis colon cancer. Nature Genet 5: 279–282
Liu B, Nicolaides N, Markowitz S, Wilson JKV, Parsons RE, Jen J, Papadopolous N, Peltomaki P, de la Chapelle A, Hamilton SR, Kinzler KW, Vogelstein B (1995) Mismatch repair gene defects in sporadic colorectal cancers with microsatellite instability. Nature Genet 9: 48–55
Loeb LA (1989) Endogeneous carcinogenesis: molecular oncology into the twenty-first century – presidential address. Cancer Res 49: 5489–5496
Loeb LA (1991) Mutator phenotype may be required for multistage carcinogenesis. Cancer Res 51: 3075–3079
Loeb LA (1994) Microsatellite instability: Marker of a mutator phenotype in cancer. Cancer Res 54: 5059–5063
Loeb LA, Cheng KC (1990) Errors in DNA synthesis: a source of spontaneous mutations. Mutat Res 238: 297–304
Loeb LA, Preston BD (1986) Mutagenesis by apurinic/apyrimidinic sites. Annu Rev Genet 20: 210–230
Loeb LA, Springgate CF, Battula N (1974) Errors in DNA replication as a basis of malignant change. Cancer Res 34: 2311–2321
Lothe RA, Peltomaki P, Meling GI, Aaltonen LA, Nystrom-Lahti M, Pylkkanen L, Heimdal L, Andersen TI, Moller P, Rognum TO, Fossa SD, Haldorsen T, Langmark F, Brogger A, de la Chapelle A, Borresen A-L (1993) Genomic instability in colorectal cancer: relationship to clinicopathological variables and family history. Cancer Res 53: 5849–5852
Low KB (1988) The recombination of genetic material. Academic, San Diego
Malkin D, Li FP, Strong LC, Fraumeni JF Jr, Nelson CE, Kim DH, Kassel J, Gryka MA, Bischoff FZ, Tainsky MA, Friend SH (1990) Germ line p53 mutations in a familial syndrome of breast cancer sarcomas and other neoplasms. Science 250: 1233–1238
Mitelman F ed (1991) Catalogue of chromosome aberrations in cancer, 4th ed. Wiley-Liss, New York
Modrich P (1987) DNA mismatch correction. Annu Rev Biochem 56: 435–466
Nowell PC (1976) The clonal evolution of tumor cell populations. Science 194: 23–28
Papadopoulos N, Nicolaides NC, Wei Y-F, Ruben SM, Carter KC, Rosen CA, Haseltine WA, Fleischmann RD, Fraser CM, Adams MD, Venter JC, Hamilton S, Petersen GM, Watson P, Lynch H, Peltomaki P, Mecklin J-P, de la Chapelle A, Kinzler KW, Vogelstein B (1994) Mutation of a mutL homolog in hereditary colon cancer. Science 263: 1625–1629
Peltomaki P, Aaltonen LA, Sistonen P, Pylkkanen L, Mecklin J-P, Jarvinen J, Green JS, Jass JR, Weber JL, Leach FS, Petersen GM, Hamilton SR, de la Chapelle A, Vogelstein B (1993) Genetic mapping of a locus predisposing to human colorectal cancer. Science 260: 810–812
Radman M, Wagner R, Kricker MC (1993) Homologous DNA interactions in the evolution of gene and chromosome structure. In: Davies KE, Warren ST (eds) Genome rearrangement and stability. Cold Spring Harbor Laboratory Press, Plainview, pp 139–152
Reid TM, Loeb LA (1993) Tandem double CC→TT mutations are produced by reactive oxygen species. Proc Natl Acad Sci USA 90: 3904–3947

Rhyu MG, Park WS, Meltzer SJ (1994) Microsatellite instability occurs frequently in human gastric carcinoma. Proc Am Assoc Cancer Res 35: 3220

Risinger JL, Berchuck A, Kohler MF, Watson P, Lynch HT, Boyd J (1993) Genetic instability of microsatellites in endometrial carcinoma. Cancer Res 53: 5100–5103

Schachman HK, Adler J, Radding CM, Lehman IR, Kornberg A (1960) Enzymatic synthesis of deoxyribonucleic acid. VII. Synthesis of a polymer of deoxyadeylate and deoxythymidylate. J Biol Chem 253: 3242–3249

Shridhar V, Siegfried J, Hunt J, del Mar Alonso M, Smith DI (1994) Genetic instability of microsatellite sequences in many non-small cell lung carcinomas. Cancer Res 54: 2084–2087

Soll DR, Morrow B, Srikantha T (1953) High-frequency phenotypic switching in Candida albicans. Trends Genet 9: 61–65

Srivastava S, Zou A, Pirollo K, Blattmer W, Chang EH (1990) Germ-line transmission of a mutated p53 gene in a cancer-prone family with Li-Fraumini syndrome. Nature 348: 747–749

Sugimura T (1992) Multistep carcinogenesis: a 1992 perspective. Science 258: 603–607

Thibodeau SN, Bren G, Schaid D (1993) Microsatellite instability in cancer of the proximal colon. Science 260: 816–819

Umar A, Boyer JC, Thomas DC, Nguyen DC, Risinger JI, Boyd J, Ionov Y, Perucho M, Kunkel TA (1994) Defective mismatch repair in extracts of colorectal and endometrial cancer cell lines exhibiting microsatellite instability. J Biol Chem 269: 14367–14370

van de Ploeg LHT, Gottesdiener K, Lee MG-S (1992) Antigenic variation in African trypanosomes. Trends Genet 8: 452–457

Vogelstein B, Fearon ER, Kern SE, Hamilton SR, Preisinger AC, Leppert M, Nakamura Y, White R, Smits AMM, Bos JL (1988) Genetic alterations during colorectal-tumor development. N Engl J Med 319: 525–532

Warren ST, Schultz RA, Chang C-C, Wade MH, Trosko JE (1981) Elevated spontaneous mutation rate in Bloom syndrome fibroblasts. Proc Natl Acad Sci USA 78: 3133–3137

Surveillance and Genome Stability in Budding Yeast: Implications for Mammalian Carcinogenesis

F. SPENCER

1 Introduction

The recognition of cancer as a genetic disease has brought with it the observation that tumor development is the outcome of multiple independent mutational events occurring during somatic tissue growth (PETO et al. 1975; FEARON and VOGELSTEIN 1990). These mutations in combination lead to abnormalities in growth control and genome instability in cancer cells. Advances in model organism studies have shown that these two attributes are frequently linked; that abnormalities in the control of cell cycle progression are often associated with increased errors in replicating and transmitting the parental genome to daughter cells. Because fundamental aspects of cell cycle control and chromosome distribution are clearly conserved in eukaryotic organisms, studies in model experimental systems are highly relevant to elucidation of the roles of these processes in tumor development in humans (HARTWELL and KASTAN 1994). This review will focus on recent advances in cell cycle control and genetic instability in the budding yeast *Saccharomyces cerevisiae*, with emphasis on experimental topics in which loss of cell cycle progression control decreases the fidelity of chromosome transmission to daughter cells. Viable single gene mutations that perturb both cell cycle control and genome stability in model organisms represent promising candidate tumor suppressor homologues. At this time, the connections between these functions in yeast and mammalian carcinogenesis are largely speculative, although the parallels are strong and warrant consideration.

Ross 1167, Center for Medical Genetics, Johns Hopkins University School of Medicine, 720 Rutland Avenue, Baltimore, MD 21205, USA

The area of chromosome transmission fidelity has made substantial recent progress in budding yeast. This is due in large part to the fact that chromosomal elements required for chromosome replication and segregation occupy short segments of DNA, and are easily manipulated in bacterial shuttle vector systems (reviewed in NEWLON 1988). This feature of yeast chromosome biology has allowed the manipulation of DNA elements in vitro, which can be subsequently reintroduced into a yeast host by DNA-mediated transformation for functional assay in vivo. In addition, the highly active homologous recombination system of budding yeast facilitates the incorporation of engineered structures into defined positions on natural yeast chromosomes for functional characterization in chromosomal context (reviewed in ROTHSTEIN 1991). These pragmatic experimental advantages have supported a rapidly advancing field of study focused on identifying determinants of chromosome transmission fidelity and characterizing their cellular roles. The current molecular genetic and biochemical analysis of proteins that function at budding yeast centromeres, origins of replication, and telomeres (reviewed in NEWLON 1988; ZAKIAN et al. 1990; HEGEMANN and FLEIG 1993; SUGINO 1995; TOYN et al. 1995) has already provided a firm foundation for the study of structural and regulatory aspects of the chromosome cycle in this organism.

The fidelity of chromosome transmission depends on the interaction of these structural elements with controlling proteins that ensure temporal order in the execution of cell division processes. The subject of cell cycle control embodies a major focus of productive research in a large number of organisms, and many molecular components of cell cycle progression control through the activity of cyclin-dependent kinases are clearly conserved among widely divergent species (reviewed in COLEMAN and DUNPHY 1994). As in all eukaryotes, activity of the yeast $p34^{CDC28}$ protein kinase catalytic subunit (the $p34^{CDC2}$ homologue) is a central element in the machinery that controls progression through mitosis in budding yeast (PIGGOTT et al. 1982; REED and WITTENBERG 1990; FITCH et al. 1992; RICHARDSON et al. 1992). $p34^{CDC28}$ activity in the G_2/M portion of the yeast cell cycle requires association with B-type cyclins, six of which are described in the literature (EPSTEIN and CROSS 1992; FITCH et al. 1992; RICHARDSON et al. 1992; SCHWOB and NASMYTH 1993). General themes in regulation of cyclin-dependent kinase activities include changes in phosphorylation status (reviewed in SOLOMON 1993) and association with inhibitory subunits (reviewed by PETER and HERSKOWITZ 1994). Studies aimed at the elucidation of the specific roles of the multiple forms of p34 kinases comprise an active area of research. The target in vivo substrates and regulation of kinase activity of these complexes are not fully described in any organism, although there is no shortage of candidates (NIGG 1993). Though much further work needs to be done in multiple organisms to distinguish the fundamentally shared from species-specific mechanisms, many observations made in model systems will be of value in probing similar functions in mammalian cells.

The cell morphology of budding yeast also provides a useful experimental tool for studies in cell cycle progression. This feature was recognized 3 decades ago by HARTWELL who used it to generate and study a large collection of cell division cycle (*cdc*) mutants (reviewed in HARTWELL 1974; PRINGLE and HARTWELL 1981). Elegant analyses of these mutants have pioneered much of what we now understand about the organization of the cell cycle. The relationship between bud morphology and cell cycle phase in *S. cerevisiae* is shown in the top portion of Fig. 1, which depicts easily scored changes in cell nuclear morphology. These morphological changes reflect progress through the four classically defined cell cycle phases G_1, S, G_2, and M. While the simplicity of the morphological assay makes it a powerful experimental tool, it is not without limitation. In particular, the transitions from S to G_2 and G_2 to M phases are not well represented by landmark morphological events. Many *cdc* mutants and many experimental treatments will cause arrest in a large budded uninucleate state, which may represent late S phase, G_2, or early mitosis, and thus other criteria must be applied to differentiate among these possibilities.

The lack of clarity in defining S to G_2 and G_2 to M phase transitions is not limited to budding yeast (nearly all experimental systems rely on methods such as flow cytometry to detect the completion of bulk DNA synthesis, and hence the S to G_2 transition), but the problem of identifying a G_2/M transition is especially acute. Hallmarks of the initiation of mitotic prophase in yeast are either not visible at this level of analysis (e.g., chromosomal condensation requires in situ hybridization for detection; Guacci et al. 1994), or do not occur (e.g., nuclear envelope dissolution is not observed in budding yeast; Byers 1981), or occur at a much earlier time in the cell cycle than in other eukaryotes (e.g., spindle pole separation and spindle microtubule formation are nearly concurrent with the initiation of DNA synthesis; Byers 1981). The majority of cell cycle studies in budding yeast have therefore relied on anaphase to indicate mitosis, or genetic arguments to differentiate preanaphase mitotic arrest from late S or G_2 arrest.

Mitosis, as a component of cell division, is an irreversible process. It has been argued that cells have evolved mechanisms that ensure readiness prior to the initiation of mitosis (e.g., Murray and Kirschner 1989; Nurse 1990; Murray 1992). However, mitosis is not a single event, but a complex series of steps (reviewed in Ault and Rieder 1994; Koshland 1994) within which there may be controls to ensure that the segregation of cellular contents will result in two viable daughters. The box associated with M phase in Fig. 1 lists events that are classically associated with mitosis in eukaryotes whether or not they have been visualized in budding yeast. These events are arbitrarily organized into three groups that can be thought of as representing entry into, execution of, and exit from mitosis. All of these events are essential for a successful division and clearly must occur in a coordinated fashion.

The number of operational control points represented within this series of events and their relationships remain to be determined. This is an important problem to be considered in the interpretation of studies addressing the nature of controlled progression through the G_2/M portion of the cycle, especially when considering results from different organisms in presumed conserved processes. The criteria used to identify cell cycle position are dictated by the landmarks available in the relevant experimental systems. Budding yeast landmarks are somewhat different from those available in mammalian cells and therefore perhaps worth brief

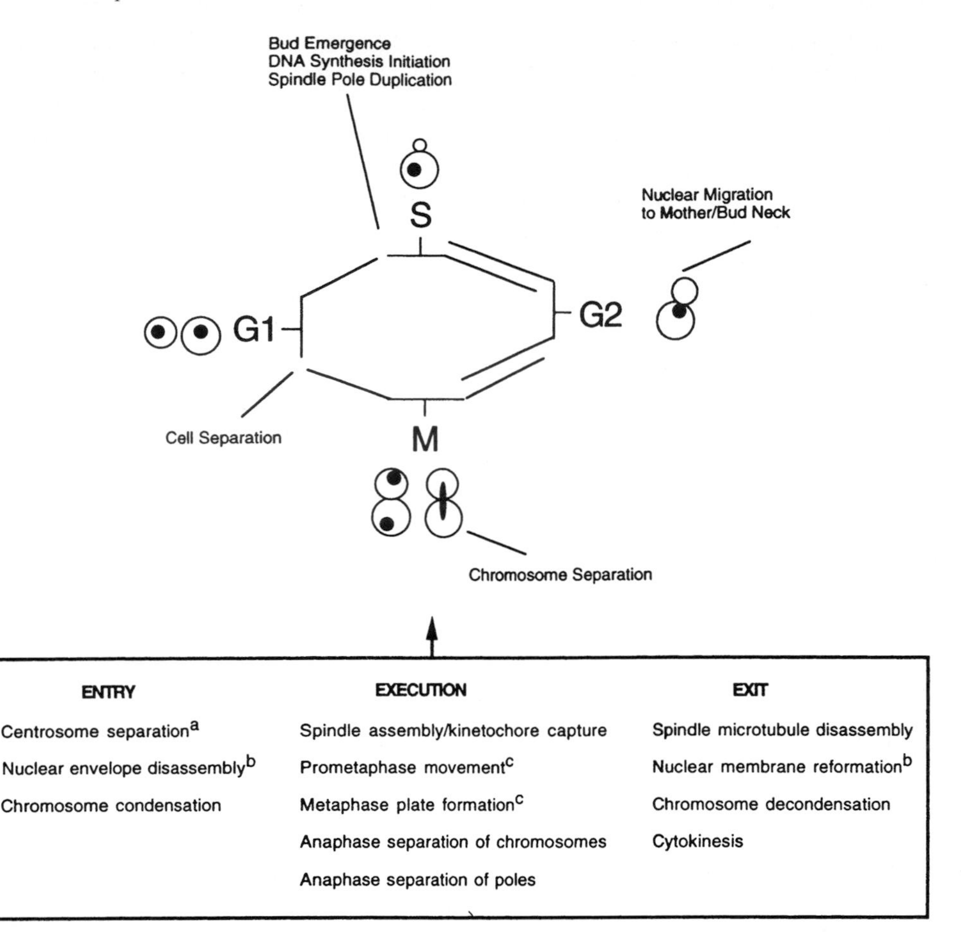

Fig. 1. Simple morphological criteria used in determining the cell cycle stage of a yeast cell. The M to G1 and G1 to S transitions can be scored using cell separation and bud formation as criteria. The S to G2 and G2 to M transitions are inferred, and the overlap between these phases as determined by cytology is indicated. The anaphase separation of chromosomes and poles is observed as simultaneous elongation of the DNA mass and spindle. It is not known if budding yeast has distinct anaphase A (chromosomal separation) and anaphase B (pole separation) movement. The *box* contains a canonical list of mitotic events whether or not they are observed in budding yeast. The large budded uninucleate cell morphology (often referred to as being in G2/M arrest or delay) encompasses late S phase, G2, and preanaphase mitotic stages. *a*, occurs early in S phase in budding yeast; *b* does not occur in budding yeast; *c*, is not cytologically distinct in budding yeast

description here (reviewed in Byers 1981). The spindle pole bodies (centrosomes of yeast) undergo separation very early in S phase, nuclear envelope disassembly is not observed, and chromosome condensation can be visualized using in situ hybridization techniques. Because yeast chromosomes contain centromeres that direct segregation of sister chromatids on a microtubule-based spindle (Peterson and Ris 1976; Byers 1981), a process in which kinetochores associate with spindle microtubules must occur. Indeed, yeast centromere DNA/protein complexes are observed to interact with microtubules in vitro (Kingsbury and Koshland 1993; Sorger et al. 1994), but when this association is first established in vivo is not known. Because individual yeast chromosomes are not delimited by light- or electron-microscopic techniques (Peterson and Ris 1976), prometaphase and metaphase states have not been distinguished. Prior to anaphase in yeast, the entire chromosome mass is seen to undergo rapid oscillatory movement (Palmer et al. 1989; Yeh et al. 1995) which may represent these phases. Anaphase itself is observed as simultaneous separation of both chromosomal masses and spindle poles in wild-type cells. In exiting mitosis, spindle microtubule disassembly occurs (Byers 1981), and chromosome decondensation is inferred (Guacci et al. 1994). Prior to or within the processes of cytokinesis, the yeast nuclear membrane must be resolved into two distinct organelles. Cytokinesis occurs at the bud neck and requires activities associated with a 10-nm filament ring (Haarer and Pringle 1987). Cell separation after cytokinesis requires chitin synthesis (reviewed in Ballou 1982).

2 Checkpoints and Surveillance: The Budding Yeast *RAD9* Example

The correct temporal order of critical events in the cell cycle is controlled by mechanisms that ensure the dependency of later processes on the completion of necessary earlier processes. This observation has been formalized and explored by Weinert and Hartwell (Weinert and Hartwell 1988; Hartwell and Weinert 1989), and termed "checkpoint control" which is mediated by "surveillance", and by Murray and Kirschner (1989), who used the term "feedback control." Experimental evidence from many systems strongly suggests widespread existence of such dependencies (Hartwell and Weinert 1989; Murray and Kirschner 1989). In defining the checkpoint concept, Hartwell and Weinert have formally defined two distinct types of dependent relationships: substrate–product limited dependency versus dependency governed by checkpoint control. In a substrate–product limited process, the rate of change is governed by the intrinsic properties of a structure, with no external regulation imposed. At a cell cycle checkpoint, however, extrinsic control governs progression via the activity of a surveillance pathway which provides a mechanism for preventing progress in a dependent process until prerequisite events have been completed. This hypothesis of checkpoint control by surveillance provides an intellectually satisfying mechanism for the temporal coordination of spatially separate events.

In practice, how can surveillance be experimentally detected? Genetic analysis of the function of the yeast *RAD9* gene provided the first compelling experimental paradigm (Weinert and Hartwell 1988). DNA damage-induced delay of mitosis required wild-type *RAD9* protein function: lack of this delay in a *rad9* deletion mutant was associated with increased chromosome loss and cell death. Restoration of delay by slowing the cell cycle with a microtubule inhibitor (provided at nonarresting concentrations) rescued viability in irradiated cells. The simplest interpretation of this result was that *rad9* deletion cells were capable of repairing radiation damage, but had lost the ability to delay cell cycle progression while the repair functions did their work. Thus, it was proposed that the wild-type *RAD9* gene operated within a surveillance system controlling the timing of cell division in response to the state of the chromosomal DNA.

This analysis established three criteria for the recognition of surveillance at a checkpoint. First, a cell cycle delay or arrest was induced by an experimental treatment causing damage (in this case, irradiation). Second, the delay or arrest was found to depend on a gene product whose removal also resulted in loss of fidelity of the relevant process (in this case, observed as chromosome loss and cell death). Third, reintroduction of a delay in cell cycle progress by an independent method (in this case, low concentration of a microtubule inhibitor) restored fidelity of the process affected in the delay-deficient mutant. In principle, these criteria can be applied to the analysis of any mutant lacking a delay or arrest response. However, the third and most convincing criterion depends on the identification of a delay-inducing treatment that is both unrelated to the surveillance pathway under study, and appropriately positioned in the cell cycle to allow the results of consequences of repair processes to be revealed. In the *RAD9* study, partial inhibition of microtubule polymerization was informative, but in other studies a similar situation could prove to be difficult to achieve.

This is not a series deterrent as limitations encountered within classical genetic approaches will be complemented by molecular genetics and biochemistry. The notion of "extrinsic control" is best modeled as an inducible reaction cascade with a diffusible or transported component. In this hypothesis, at a minimum there must be an inducer, a signal (including the diffusible or transported component), and one or more effectors, the timing of whose substrate–product conversion is altered by input from the signal. Considering cumulative experience in the analysis of signal transduction pathways, it is highly likely that molecular and biochemical approaches will link legitimate components identified by genetic studies.

3 Budding Yeast Genes That Monitor Integrity of the Chromosomal DNA

Further screening of yeast *rad* genes identified two additional genetic loci, *RAD17* and *RAD24*, with properties similar to *RAD9* (Weinert and Hartwell 1993; Weinert et al. 1994). In a study designed to characterize the specificity of the *RAD9* checkpoint, *cdc* mutants encoding known replication enzymes were found to depend on *RAD9* for their G_2/M arrests and for their recovery on return to permissive temperature (Weinert and Hartwell 1993). The "rapid death" phenotype observed in replication mutants that lacked *RAD9* was exploited in the design of a primary screen for G_2/M surveillance genes. This screen identified three *MEC* (mitosis entry checkpoint) loci: *MEC1/ESR1, MEC2/SAD1/RAD53/SPK1*, and *MEC3* (Weinert et al 1994; Allen et al. 1994; Zheng et al. 1993), bringing the total number of published genes with roles in DNA damage surveillance to six.

The primary amino acid sequence of three of these genes have been reported. *RAD9* exhibits no obvious homology to known proteins (Weinert and Hartwell 1990). *MEC2* encodes an essential dual-specificity protein kinase which is also required for a DNA damage-induced transcriptional response (Zheng et al. 1993; Allen et al. 1994). The *MEC1* predicted protein sequence exhibits significant similarity to the budding yeast *TEL1* gene product as well as the recently cloned human gene for ataxia telangiectasia, *ATM* (Greenwell et al. 1995; Morrow et al. 1995; Savitsky et al. 1995). These protein sequences provide additional members to a group with homology to phosphatidylinositol-3 kinases, although biochemical activities of *MEC1, TEL1*, and *ATM* have not been characterized. Abnormalities in human ataxia telangiectasia are consistent with a DNA damage checkpoint defect, and affected individuals are at increased risk of tumor development (reviewed by Shiloh 1995). It has been hypothesized that human *ATM* functions are shared by activities of *TEL1* and *MEC1* proteins in yeast (Greenwell et al. 1995; Morrow et al. 1995). Protein and functional homology are also evident in analysis of a fruit fly homolog (*mei-41*, Hari et al. 1995) and a fission yeast homolog (*rad3*; Seaton et al. 1992). Interestingly, AT cells show defects in p53 induction by DNA-damaging agents (Kastan et al. 1992). This observation suggests that the mammalian p53 damage response includes surveillance elements homologous to the budding yeast DNA damage response. Further work on these related genes holds promise for determining what aspects are conserved within the cell physiology of DNA damage surveillance pathways in eukaryotes.

In the pioneering study that defined the *RAD9*-dependent DNA damage checkpoint in G_2/M, logarithmically growing cell populations were subjected to DNA-damaging treatments and found to accumulate late in the cell cycle, prior to anaphase of mitosis. Subsequent studies showed that synchronous yeast cell populations irradiated in G_1 exhibit a pronounced delay prior to entry into S phase, and that this S initiation delay depends on *RAD9, RAD24*, and *MEC2/RAD53* (Allen et al. 1994; Siede et al. 1993, 1994). The fact that all three genes tested (of the six known to operate at G_2/M) are required for S phase initiation delay function suggests that a single surveillance pathway can operate to arrest cells in response to DNA damage at these two cell cycle stages.

However, the pathway appears to be branched: a subset of DNA damage response loci are required for the maintenance of cell cycle arrest in the presence of a DNA replication block at initiation of DNA synthesis (Weinert et al. 1994). All six *MEC* and *RAD* surveillance loci were identified by their requirement for response to DNA insults that directly alter DNA structure or prevent the completion of DNA replication. When these mutants were tested for arrest in the presence of hydroxyurea, two different responses were observed: *mec3, rad9, rad17*, and *rad24* mutants arrested their cell cycles, whereas *mec1* and *mec2/rad53* mutants failed to arrest and died rapidly. The effect of hydroxyurea treatment is to block nucleotide synthesis by inhibiting ribonucleotide reductase, and thus prevent DNA replication once the internal cellular pools are drained. This experiment suggests that the physiological response to unreplicated DNA is qualitatively different from that to damaged DNA and utilizes an overlapping but not identical control pathway.

Interestingly, surveillance pathway function is also implicated in the control of the rate of passage through S phase. In a recent study, it was observed that alkylation damage to DNA (by exposure to methylmethane sulfonate) slows progress through S phase, as measured by flow cytometric analysis, and that this slowed pace requires the *MEC1* and *RAD53* gene functions (PAULOVICH and HARTWELL 1995). This observation strongly suggests that at least some surveillance pathway functions can direct modification of the DNA replication machinery or chromosomal proteins that modulate the rate of DNA replication. The relationship between this type of progression control and those which appear to delay cell cycle phase transitions (G_1 to S and G_2 to M) is not yet apparent.

It seems likely that related surveillance pathways that monitor DNA integrity are capable of responding to a variety of abnormal structures within chromosomes. Interestingly, NAVAS et al. (1995) have recently demonstrated that the C-terminal tail of budding yeast DNA polymerase epsilon is required for the DNA damage transcriptional response in S phase and for S phase arrest in response to hydroxyurea. These results suggest that the activities of DNA polymerase epsilon are intimately associated with a DNA damage-sensing mechanism.

The relationship between DNA damage pathway control and cell cycle progression control via $p34^{CDC28}$ kinase activity is open to speculation. Although cyclin-dependent kinases are universal cell cycle regulators (NURSE 1990; COLEMAN and DUNPHY 1994), their molecular physiology is not simple. Experiments in fission yeast and frogs have shown that an inhibitory phosphorylation on tyrosine 15 of the cyclin-dependent kinase catalytic subunit was required for cell cycle arrest in the presence of incomplete DNA replication or DNA damage (ENOCH and NURSE 1990; SMYTHE and NEWPORT 1992). Though an analogous modification clearly occurs on tyrosine 19 of $p34^{CDC28}$ in budding yeast, the functional connection with surveillance of DNA integrity appears not to be conserved in *S. cerevisiae* (AMON et al. 1992; SORGER and MURRAY 1992; STUELAND et al. 1993). As these experiments would have been most sensitive to the G_2/M block, it is clear that G_2/M activity of the *RAD9* pathway does not act via Tyr 19 phosphorylation. Clearly, however, this result does not preclude a role for $p34^{CDC28}$ interaction.

4 A Surveillance Pathway Monitors Integrity of the Mitotic Spindle in Budding Yeast

There is very strong evidence for a widely conserved checkpoint that responds to the state of the mitotic spindle. Drugs that inhibit the formation of spindle microtubules, such as benzidimazoles, induce an arrest in mitosis in a large number of diverse organisms (reviewed in RIEDER and PALAZZO 1992). Under conditions that completely disrupt spindle assembly, this arrest occurs in prometaphase (as indeed there are no microtubules to interact with the chromosomes in the formation of a metaphase plate). Under conditions that interfere with but do not destroy microtubule formation, cells are observed to arrest in metaphase. Whether these are physiologically distinct arrest pathways remains an open question. The duration of drug-induced arrest varies with drug concentration, organism, and cell type (KUNG et al. 1990; RIEDER and PALAZZO 1992; SCHIMKE et al. 1994), but the presence of a pause is essentially universal, suggesting a fundamental and conserved mechanism.

Preanaphase arrest is also associated with abnormal kinetochore configuration in diverse experimental paradigms and in many distantly related eukaryotes. Methods providing experimental interference that lead to preanaphase arrest or delay include breakage of spindle fibers with a glass needle in grasshopper cells (NICKLAS 1967), breakage of spindle fibers using a UV microbeam in newt cells (ZIRKLE 1970), and microinjection of antibodies that bind chromosomal centromeres or kinetochores in cultured human (BERNAT et al. 1990; YEN et al. 1991; TOMKIEL et al. 1994; CAMPBELL and GORBSKY 1995) and mouse (SIMERLY et al. 1990) cells. Although these experiments point to a preanaphase signal that blocks chromosome segregation, they themselves do not differentiate between an arrest resulting from a steric mechanism (i.e., a substrate–product relationship as discussed above) versus one requiring extrinsic control by a surveillance pathway. The strongest evidence for the activity of a surveillance pathway is provided by studies in human PtK cells (RIEDER et al. 1994) and budding yeast (SPENCER and HIETER 1992). In human PtK cells, the timing of coordinate anaphase separation of all chromosomes is markedly delayed in cells in which a single chromosome spontaneously fails to achieve spindle attachment by the time the other chromosomes are aligned at the metaphase plate. Moreover, the time interval between bipolar attachment of all chromosomes and anaphase initiation is the same in delayed or nondelayed mitoses. This is consistent with the hypothesis that anaphase chromatid separation is extrinsically controlled by a signal that communicates the unready state of any unattacked kinetochore and that, when metaphase spindle assembly is complete, a stereotypical set of biochemical events is initiated that releases the sister chromatids and segregates them to their respective poles. Recent data from micromanipulation experiments in praying mantis spermatocytes (LI and NICKLAS 1995) strongly suggest that tension on the chromosomal kinetochores, rather than physical association between kinetochore and pole, is the important feature of bipolar attachment. The description of the differential expression of a kinetochore phosphoepitope may provide a clue to structural differences between attached and unattached kinetochores at the molecular level. This epitope is apparent in prophase on all kinetochores and absent in metaphase except on the kinetochores of unattached chromosomes (GORBSKY and RICKETTS 1993).

Strong evidence of a similar preanaphase delay in budding yeast benefits from recent work identifying the centromere DNA elements and many associated proteins essential for chromosome segregation (reviewed in HEGEMANN and FLEIG 1993; KILMARTIN 1994). The cloning and characterization of yeast centromere DNA sequences have revealed a 125 nucleotide conserved region composed of three elements, referred to as centromere DNA elements (CDEs). CDEI and CDEIII have palindromic character, and proteins that bind these sequences have been identified by both biochemical and genetic criteria. CDEIII is clearly essential for centromere function as point mutations central to the palindrome decrease chromosome segregation function by three orders of magnitude (HEGEMANN et al. 1988; JEHN et al. 1991). A 240-kilodalton multisubunit complex of proteins that binds CDEIII in vitro has been described (LECHNER and CARBON 1991). The three major protein species present in the complex are encoded by the kinetochore protein genes *NDC10, CTF13*, and *CEP3* (DOHENEY et al. 1993; GOH and KILMARTIN 1993; JIANG et al. 1993; STRUNNIKOV et al. 1995). Three groups have described a helix-loop-helix protein that binds the CDEI consensus as a dimer (BAKER and MASSION 1990; CAI and DAVIS 1990; MELLOR et al. 1990). This protein is encoded by a nonessential gene (*CPF1/CBF1/CEP1)* whose deletion results in a ten fold decrease in chromosome transmission fidelity. CDEII is an A+T rich region (exhibiting little other sequence conservation among centromere DNA sequences) whose function is unknown. CDEII provides an essential activity as CDEIII alone does not confer stable chromosome transmission (PANZERI et al. 1985).

The terms "kinetochore" and "centromere" have evolved highly specific, but different, meanings among groups of scientists working in different systems. A mammalian kinetochore can be visualized in electron micrographs and is defined as a multilayered structure associated with the primary constriction of mitotic chromosome (RIEDER 1982; EARNSHAW 1991). In budding yeast, the chromosomes are not cytologically distinct structures, and thus there is no visible constriction to which this term could refer. For the purposes of this review, the term yeast "kinetochore" will refer to the entire DNA/protein complex responsible for chromosome attachment to the mitotic spindle, and chromosome movement to the poles in mitosis.

Preanaphase delay induced by budding yeast kinetochore alteration has been detected in temperature-sensitive mutants of *NDC10, CTF13,* and *CEP3* (DOHENEY et al. 1993; GOH and KILMARTIN 1993; STRUNNIKOV et al. 1995). The delay is observed as accumulation of large budded uninucleate cells in culture at semipermissive or nonpermissive temperatures. To date, the period that is elongated prior to anaphase remains largely uncharacterized, but is generally assumed to occur in G_2 or early M. Preanaphase delay is also observed in yeast cells containing a single test chromosome with a centromere DNA mutation (SPENCER and HIETER 1992). Interestingly, in these experiments only some centromere DNA mutations caused delay, and the extent of delay did not correlate with the degree of chromosome missegregation. This observation indicates that kinetochore abnormalities that compromise chromosome segregation differ in kind. The ability of a single abnormal kinetochore to induce preanaphase delay strongly supports the existence of a preanaphase surveillance pathway in budding yeast and suggests that mammalian and budding yeast are likely to share conserved elements in progression control through mitosis.

Recently, two groups have identified mutant strains that bypass arrest or delay in the presence of the microtubule-inhibiting drug benomyl. The *bub* (budding uninhibited by benzidimazole) mutants (HOYT et al. 1991) were isolated for two diagnostic phenotypes: they are unable to recover from brief exposure to arresting concentrations of drug (sufficient to completely prevent the formation of a spindle), and they undergo unscheduled bud formation in the absence of nuclear division. The *mad* (mitotic arrest defective) mutants (LI and MURRAY 1991) were identified by decreased viability in the presence of low (nonarresting) concentrations of drug and an inability to slow cell division time under these conditions. The *bub* mutant screen yielded three mutants defining two genetic loci, *BUB1* and *BUB2*. During the cloning of the *BUB1* gene, *BUB3* was discovered as a low copy suppressor of *bub1-1*. The *mad* screen generated three mutants defining three loci: *MAD1, MAD2,* and *MAD3*. Neither screen was performed to saturation, and there are therefore likely to be many more genetic loci that could be identified using this approach.

Recent data (PANGILINAN and SPENCER, 1996) indicate that the components of the spindle surveillance pathway active in the presence of antimicrotubulte drugs are also required for the delay induced by mutant kinetochores in yeast. In *bub1* and *mad2* mutants, single chromosomes containing a delaying centromere DNA mutation are specifically and profoundly destabilized. This result indicates not only that *BUB1* and *MAD2* are required for the cell cycle delay, but that the delay is important for mitotic segregation of the mutant kinetochore. Further, the delay induced by the temperature-sensitive kinetochore protein mutant *ctf13-30* is abolished in *bub1* and *mad2* mutants. These data are consistent with the hypothesis that the kinetochores are inducers of a spindle surveillance pathway, although there is no evidence to suggest that kinetochores are the only inducers. It is also not clear what abnormal kinetochore property is important for delay induction although, by analogy with the observations in mammalian cells, it is tempting to speculate on the existence of a conserved response to microtubule attachment defect.

The analysis of spindle surveillance at the molecular level is just beginning. The predicted amino acid sequences for *BUB1, BUB2, BUB3, MAD1*, and *MAD2* have been determined (HOYT et al. 1991; LI and MURRAY 1991; and the public databases). Unfortunately, the primary sequences of four of these (*BUB2, BUB3, MAD1* and *MAD2*) do not predict known molecular activities. *MAD2* was originally reported to encode a prenyl transferase; it has since been discovered that the cell cycle control function isolated in the original mutant is encoded by the neighboring open reading frame (LI et al. 1994). Interestingly, *BUB1* encodes a dual specificity protein kinase (ROBERTS et al. 1994). It has been speculated that *BUB3* may encode a regulator of the *BUB1* kinase: the *BUB3* gene was originally isolated as a low copy suppressor of *bub1-1*, *BUB3* coimmunoprecipitates with *BUB1*, and *BUB3* is phosphorylated by *BUB1* in vitro (ROBERTS et al. 1994).

DNA damage surveillance and spindle surveillance impose arrests in the G_2/M portion of the cell cycle. Do these surveillance pathways intersect? It is clear that *rad9* mutants delay in the presence of benomyl (WEINERT and HARTWELL 1988) and do not play a role in the yeast kinetochore-induced delay (PANGILINAN and SPENCER 1996). Moreover, the three *mad* mutants regain benomyl resistance in the presence of hydroxyurea, suggesting that a DNA damage surveillance pathway is operative (Li and Murray 1991). These observations argue that the surveillance pathways are distinct. However, a large body of evidence indicates that in normal cells the execution of mitosis depends on prior DNA replication, and in fission yeast and frogs this linkage involves control of cell cycle progression through $p34^{CDC2}$ (reviewed in Nurse 1990; Murray 1992). The relationship between cyclin-dependent kinase cell cycle control and the surveillance pathways monitoring DNA integrity and spindle structure has not been determined.

An intriguing recent observation from the study of mammalian p53 suggests a relationship between DNA damage and spindle structure surveillance pathways. Although the S phase-related functions of p53 are best characterized, the behaviour of cells from p53-deficient mice are consistent with an additional functional role in responding to simple damage at G_2/M (Cross et al. 1995). In these experiments, mouse embryonic fibroblasts lacking p53 (homozygous nulls) treated with nocodazole (an antimicrotubule drug) fail to accumulate in metaphase and rapidly generate 8N and 16N subpopulations. Polyploidy was also seen to develop spontaneously in vivo. These observations on p53 null mutants suggest that searches for common elements between DNA damage and spindle structure surveillance pathways may prove to be fruitful.

5 A Surveillance Pathway May Monitor Budding Yeast Cell Morphogenesis

A recent study in budding yeast suggests that temporal control of mitosis may be sensitive to timely cytoskeletal rearrangement (Lew and Reed 1995). Bud initiation is associated with a polarization of the actin cytoskeleton and delivery of cytoplasmic materials to the growing bud site. These events do not occur in *cdc24* and *cdc42* mutants at nonpermissive temperature, even though DNA synthesis proceeds and microtubules form a spindle. Anaphase timing under these conditions occurs $\geq$2h later than in control cells, timing that is essentially a full generation behind schedule.

Evidence that this delay may be due to the activity of an extrinsic control is found in the role played by $p34^{CDC28}$ (Lew and Reed 1995). In *cdc24* and *cdc42* cells at nonpermissive temperature, the delay observed is associated with increased phosphorylation of tyrosine 19 of $p34^{CDC28}$. Moreover, a phenylalanine substitution mutation (which cannot be phosphorylated) in $p34^{CDC28}$ substantially decreases the preanaphase delay in *cdc24* and *cdc42* mutants. Consistent with this observation, overexpression of the tyrosine phosphatase M1H1 (the *S. cerevisiae* homolog of the *S. pombe* CDC25 phosphatase that functions in controlling $p34^{CDC2}$ activity) partially abrogates the delay.

It is worth pointing out that this study presents experimental evidence for surveillance by a fundamentally different approach than the identification of a control pathway typified by the *RAD9* mutational analysis. In the LEW and REED study, an alteration of $p34^{CDC28}$ activity implies the presence of a surveillance pathway which acts directly on the controlling M phase kinase and represents the first such regulation through $p34^{CDC28}$ demonstrated in budding yeast. Although cytoskeletal defects are the most obvious morphological consequences of loss of *CDC42* and *CDC24* function, further work is needed to identify the molecular nature of the "damage" that is sensed by the cell: *CDC42* and *CDC24* encode a rho-like GTPase and interacting nucleotide exchange factor, respectively. Might there be a related surveillance activity in mammalian cells? It is a tempting speculation: in mammals, the rho-related proteins direct actin reorganization and associate with protooncogene products (reviewed in RIDLEY 1995).

Experimental evidence demonstrating the involvement of $p34^{CDC28}$ in extrinsic control of anaphase timing in yeast is particularly intriguing in light of the several lines of evidence pointing toward a role for altered p34 control in tumor formation in mammals. GALAKTIONOV et al. (1995b) have recently shown that human *CDC25* overexpression in rodent cells (predicted to promote tyrosine 15 dephosphorylation of cyclin-dependent kinases) cooperated with Ha-$RASG_12V$ or Rb loss in oncogenic focus formation, and that human *CDC25B* was overexpressed in 32% of human breast cancers tested (GALAKTIONOV et al. 1995b). Furthermore, human *CDC25* has been observed to associate with Raf1, a Ras-associated guanine nucleotide binding protein (GALAKTIONOV et al. 1995a). It will be interesting to learn whether one or more species of cyclin-dependent kinase is important for tumor development in humans, and which if any of the candidate cellular processes appear to be governed by altered control.

6 Concluding Remarks

Proteins that function to ensure high-fidelity chromosome replication and transmission present promising candidates for proteins with roles in tumor suppression in mammalian systems. Those that are not essential for cell viability, but are important for high-fidelity execution of chromosomal events, provide especially attractive candidates. In budding yeast, the pathways that provide surveillance of DNA and spindle integrity are clearly important for genome stability. Chromosome loss rates are elevated in the range of 10- to 50-fold in *rad9*, *mad1*, *mad2*, *mad3*, *bub1*, and *bub3* mutants (WEINERT and HARTWELL 1988; LI and MURRAY 1991; PANGILINAN and SPENCER 1996), indicating that surveillance activities are utilized in unperturbed cell populations. The current repertoire of surveillance genes provides a rich starting point, with the promise of additional genes to emerge from related work in the near future.

Although parallels in cell physiology indicate conservation of fundamental processes, the protein sequences of homologous genes provide the molecular links between related studies in yeast cells and mammals. The first such link between surveillance pathways under study in budding yeast and mammalian cells has recently been found in the structural and functional ATM/MEC1/TEL1 homologies. This observation provides material for the design of direct experiments to evaluate the degree of conservation of protein and pathway function in two very distantly related organisms.

The recent formulation of the concept of surveillance has affected the way in which many experiments in cell cycle control and genome stability are designed and interpreted. At this stage, there are clearly more questions about mechanism than unifying hypotheses. Happily, the armament of experimental tools available in model organisms and mammals is undergoing rapid expansion. While the cytological description of the behavior of chromosomes throughout the cell cycle is vastly superior in animal systems (especially for the study of events in mitosis), the study of homologous phenomena in yeast opens avenues to efficient molecular genetic manipulations. The utilization of the strengths of both yeast and mammalian experimental systems will greatly facilitate progress in understanding the underlying molecular mechanisms underlying surveillance.

References

Allen J, Zhou Z, Siede W, Friedberg E, Elledge S (1994) The SAD1/RAD53 protein kinase controls multiple checkpoints and DNA damage-induced transcription in yeast. Genes Dev 8: 2416–2428

Amon A, Surana U, Muroff I, Nasmyth K (1992) Regulation of $p34^{CDC28}$ tyrosine phosphorylation is not required for entry into mitosis in S. cerevisiae. Nature 355: 368–371

Ault J, Rieder C (1994) Centrosome and kinetochore movement during mitosis. Curr Biol 6: 41–49

Baker R, Massison D (1990) Isolation of the gene encoding the Saccharomyces cerevisiae centromere binding protein CP1. Mol Cell Biol 10: 2458–2467

Ballou C (1982) Yeast cell wall and cell surface. In: Strathern J, Jones E, Broach J (eds) The molecular biology of the yeast saccharomyces: metabolism and gene expression. Cold Spring Harbor, New York, pp 335–360

Bernat R, Borisy G, Rothfield N, Earnshaw WC (1990) Injection of anti-centromere antibodies in interphase disrupts events required for chromosome movement in mitosis. J Cell Biol 111: 1519–1533

Byers B (1981) Cytology of the yeast life cycle. In: Strathern J, Jones E, Broach J (eds) The molecular biology of the yeast saccharomyces: life cycle and inheritance. Cold Spring Harbor, New York, pp 59–96

Cai M, Davis R (1990) Yeast centromere binding protein CBF1, of the helix-loop-helix protein family, is required for chromosome stability and methionine prototrophy. Cell 61: 437–446

Campbell MS, Gorbsky GJ (1995) Microinjection of mitotic cells with the 3F3/2 anti-phosphoepitope antibody delays the onset of anaphase. J Cell Biol 129: 1195–1204

Carr A, Hoekstra M (1995) The cellular responses to DNA damage. Trends Cell Biol 5: 32–40

Coleman TR, Dunphy WG (1994) Cdc2 regulatory factors. Curr Opin Cell Biol 6: 877–882

Cross S, Sanchez D, Morgan C, Schimke M, Ramel S, Idzerda R, Raskind W, Reid B (1995) A p53-dependent mouse spindle checkpoint. Science 267: 1353–1356

Doheny K, Sorger P, Hyman A, Tugendriech S, Spencer F, Hieter P (1993) Identification of essential components of the S. cerevisiae kinetochore. Cell 73: 1–14

Earnshaw WC (1991) When is a centromere not a kinetochore? J Cell Sci 99: 1–4

Enoch T, Nurse P (1990) Mutation of fission yeast cell cycle control genes abolishes dependence of mitosis on DNA replication. Cell 60: 665–673

Epstein C, Cross F (1992) CLB5: a novel B cyclin from budding yeast with a role in S phase. Genes Dev 6: 1695–1706

Fearon E, Vogelstein B (1990) A genetic model for colorectal tumorigenesis. Cell 61: 759–767

Fitch I, Dahmann C, Surana U, Amon A, Nasmyth K, Goetsch L, Byers B, Futcher B (1992) Characterization of four B-type cyclin genes of the budding yeast Saccharomyces cerevisiae. Mol Biol Cell 3: 805–818
Ford J, al-Khodairy F, Fotou E, Sheldrick K, Griffiths D, Carr A (1994) 14-3-3 homologs required for the DNA damage checkpoint in fission yeast. Science 265: 533–535
Galaktionov K, Jessus C, Beach D (1995a) Raf1 interaction with Cdc25 phosphatase ties mitogenic signal transduction to cell cycle activation. Genes Dev 9: 1046–1058
Galaktionov K, Lee A, Eckstein J, Draetta G, Meckler J, Loda M, Beach D (1995b) CDC25 phosphatases as potential human oncogenes. Science 269: 1575–1577
Goh P, Kilmartin J (1993) NDC10: a gene involved in chromosome segregation in Saccharomyces cerevisiae. J Cell Biol 121: 503–512
Gorbsky G, Ricketts W (1993) Differential expression of a phosphoepitope at the kinetochores of moving chromosomes. J Cell Biol 122: 1311–1321
Greenwell P, Kronmal S, Porter S, Gassenhuber J, Obermaier B, Petes T (1995) TEL1, a gene involved in controlling telomere length in S. cerevisiae, is homologous to the human ataxia telangiectasia gene. Cell 82: 823–829
Guacci V, Hogan E, Koshland D (1994) Chromosome condensation and sister chromatid pairing in budding yeast. J Cell Biol 125: 517–530
Haarer B, Pringle J (1987) Immunofluorescence localization of the Saccharomyces cerevisiae CDC12 gene product to the vicinity of the 10-nm filaments in the mother-bud neck. Mol Cell Biol 7: 3678–3687
Hari K, Santerre A, Sekelsky J, Mckim K, Boyd J, Hawley S (1995) The mei-41 gene of D. melanogaster is a structural and functional homolog of the human ataxia telangiectasia gene. Cell 82: 815–821
Hartwell L (1974) Saccharomyces cerevisiae cell cycle. Bacteriol Rev 38: 164–198
Hartwell L, Kastan M (1994) Cell cycle control and cancer. Science 266: 1821–1828
Hartwell L, Weinert T (1989) Checkpoints: controls that ensure the order of cell cycle events. Science 240: 629–634
Hegemann JH, Fleig U (1993) The centromere of budding yeast. Bioessays 15: 451–460
Hegemann JH, Shero JH, Cottarel G, Philippsen P, Hieter P (1988) Mutational analysis of centromere DNA from chromosome VI of Saccharomyces cerevisiae. Mol Cell Biol 8: 2523–2535
Hoyt A, Totis L, Roberts T (1991) S. cerevisiae genes required for cell cycle arrest in response to loss of microtubule function. Cell 66: 507–517
Jehn B, Niedenthal R, Hegemann J (1991) In vivo analysis of the Saccharomyces cerevisiae centromere CDEIII sequence: requirements for mitotic chromosome segregation. Mol Cell Biol 11: 5212–5221
Jiang W, Lechner J, Carbon J (1993) Isolation and characterization of a gene (CBF2) specifying a protein component of the budding yeast kinetochore. J Cell Biol 121: 513–519
Kastan M, Zhan Q, El-Diery W, Carrier F, Jacks T, Walsh W, Plunkett B, Vogelstein B, Fornace J (1992) A mammalian cell cycle checkpoint pathway utilizing p53 and GADD45 is defective in ataxia telangiectasia. Cell 71: 587–597
Kilmartin J (1994) Genetic and biochemical approaches to spindle function and chromosome segregation in eukaryotic organisms. Curr Opin Cell Biol 6: 50–54
Kingsbury J and Koshland D (1993) Centromere function on minichromosomes from budding yeast. Mol Biol Cell 4: 859–870
Koshland D (1994) Mitosis: back to the basics. Cell 77: 951–954
Kung A, Sherwood S, Schimke R (1990) Cell line-specific differences in the control of cell cycle progression in the absence of mitosis. Proc Natl Acad Sci USA 87: 9553–9557
Lechner J, Carbon J (1991) A 240kD multisubunit protein complex, CBF3, is a major component of the budding yeast centromere. Cell 64: 717–725
Lew D, Reed S (1995) A cell cycle checkpoint monitors cell morphogenesis in budding yeast. J Cell Biol 129: 739–749
Li R, Murray A (1991) Feedback control of mitosis in budding yeast. Cell 66: 519–531
Li R, Havel C, Watson J, Murray A (1994) The mitotic feedback control gene MAD2 encodes the alpha subunit of a prenyltransferase: correction. Nature 371: 438
Li X, Nicklas RB (1995) Mitotic forces control a cell-cycle checkpoint. Nature 373: 630–632
Mellor J, Jiang W, Funk M, Rathjen J, Barnes C, Hiz T, Hegemann J, Philippsen P (1990) CPF1, a yeast protein which functions in centromeres and promoters. EMBO J 9: 4017–4026
Morrow D, Tagle D, Shiloh Y, Collins F, Hieter P (1995) TEL1, an S. cerevisiae homolog of the gene mutated in ataxia telangiectasia, is functionally related to the yeast checkpoint gene MEC1. Cell 82: 831–840
Murray A (1992) Creative blocks: cell-cycle checkpoints and feedback controls. Nature 359: 599–604

Murray A, Kirschner MW (1989) Dominoes and clocks: the union of two views of the cell cycle. Science 240: 614–621
Navas TA, Zhou Z, Elledge SJ (1995) DNA polymerase epsilon links the DNA replication machinery to the S phase checkpoint. Cell 80: 29–39
Newlon C (1988) Yeast chromosome replication and segregation. Microbiol Rev 52: 568–601
Nicklas B (1967) Chromosome micromanipulation: induced reorientation and the experimental control of segregation in meiosis. Chromosoma 21: 17–50
Nigg E (1993) Targets of cyclin-dependent protein kinases. Curr Biol 5: 187–193
Nurse P (1990) Universal control mechanism regulating onset of M phase. Nature 344: 503–508
Palmer R, Koval M, Koshland D (1989) The dynamics of chromosome movement in the budding yeast Saccharomyces cerevisiae. J Cell Biol 109: 3355–3366
Pangilinan F, Spencer F, (1996) Abnormal kinetochore structure activates the spindle assembly checkpoint in budding yeast. Mol Biol Cell
Panzeri L, Landonio L, Sotz A, Philippsen P (1985) Role of conserved sequence elements in yeast centromere DNA. EMBO J 4: 1867–1874
Paulovich A, Hartwell L (1995) A checkpoint regulates the rate of progression through S phase in S. cerevisiae in response to DNA damage. Cell 82: 841–847
Peter M, Herskowitz I (1994) Joining the complex: cyclin-dependent kinase inhibitory proteins and the cell cycle. Cell 79: 181–184
Peterson J, Ris H (1976) Electron microscopic study of the spindle and chromosome movement in the yeast Saccharomyces cerevisiae. J Cell Sci 22: 219–242
Peto R, Roe FJ, Lee PN, Levy L, Clack J (1975) Cancer and ageing in mice and men. Br J Cancer 32: 411–426
Piggott J, Rai R, Carter B (1982) A bifunctional gene product involved in two phases of the yeast cell cycle. Nature 298: 391–393
Pringle J, Hartwell L (1981) The Saccharomyces cerevisiae cell cycle. In: Strathern J, Jones E, Broach J (eds) The molecular biology of the yeast saccharomyces: life cycle and inheritance. Cold Spring Harbor Laboratory Press, Cold Spring Harbor, pp 97–142
Reed SI, Wittenburg C (1990) Mitotic role for the Cdc28 protein kinase of Saccharomyces cerevisiae. Proc Natl Acad Sci USA 87: 5697–5701
Richardson H, Lew D, Henze M, Sugimoto K, Reed S (1992) Cyclin-B homologs in Saccharomyces cerevisiae function in S phase and in G_2. Genes Dev 6: 2021–2034
Ridley A (1995) Rho-related proteins: actin cytoskeleton and cell cycle. Curr Opin Genet Dev 5: 24–30
Rieder C (1982) The formation, structure, and composition of the mammalian kinetochore fiber. Int Rev Cytol 79: 1–58
Rieder C, Palazzo R (1992) Colcemid and the mitotic cell cycle. J Cell Sci 102: 387–392
Rieder C, Schultz A, Cole R, Sluder G (1994) Anaphase onset in vertebrate somatic cells is controlled by a checkpoint that monitors sister kinetochore attachment to the spindle. J Cell Biol 127: 1301–1310
Roberts T, Farr K, Hoyt A (1994) The Saccharomyces cerevisiae checkpoint gene BUB1 encodes a novel protein kinase. Mol Cell Biol 14: 8282–8291
Rothstein R (1991) Targeting, disruption, replacement, and allele rescue: integrative DNA transformation in yeast. Methods Enzymol 194: 281–301
Savitsky K, Bar-Shira A, Gilad S, Rotman G, Ziv Y, Vanagaite L, Tagle D, Smith S, Uziel T, Sfez S, Ashkenazi M, Pecker I, Frydman M, Harnik R, Patanjali S, Simmons A, Clines G, Sartiel A, Gatti R, Chessa L, Sanal O, Lavin M, Jaspers N, Malcolm A, Taylor R, Arlett C, Miki T, Weissman S, Lovett M, Collins F, Shiloh Y (1995) A single ataxia telangiectasia gene with a product similar to PI-3 kinase. Science 268: 1749–1753
Schimke R, Kung A, Sherwood S, Sheridan J, Sharma R (1994) Life, death, and genomic change in perturbed cell cycles. Philos Trans R Soc Lond [B] 345: 311–317
Schwob E, Nasmyth K (1993) CLB5 and CLB6, a new pair of B cyclins involved in DNA replication in Saccharomyces cerevisiae. Genes Dev 7: 1160–1175
Seaton B, Yucel J, Sunnerhagen P, Subramani S (1992) Isolation and characterization of the Schizosaccharomyces pombe rad3 gene, involved in the DNA damage and DNA synthesis checkpoints. Gene 119: 83–89
Shiloh Y (1995) Ataxia-telangiectasia: closer to unravelling the mystery. Eur J Hum Genet 3: 116–138
Siede W, Friedberg AS, Friedberg EC (1993) RAD9-dependent G_1 arrest defines a second checkpoint for damaged DNA in the cell cycle of Saccharomyces cerevisiae. Proc Natl Acad Sci USA 90: 7985–7989
Siede W, Friedberg AS, Dianova I, Friedberg EC (1994) Characterization of G_1 checkpoint control in the yeast Saccharomyces cerevisiae following exposure to DNA-damaging agents. Genetics 138: 271–281

Simerly C, Balczon R, Brinkley B, Schatten G (1990) Microinjected kinetochore antibodies interfere with chromosome movement in meiotic and mitotic mouse oocytes. J Cell Biol 111: 1491–1504
Smythe C, Newport J (1992) Coupling of mitosis to the completion of S phase in Xenopus occurs via modulation of the tyrosine kinase that phosphorylates $p34^{CDC2}$. Cell 68: 787–797
Solomon M (1993) Activation of the various cyclin/cdc2 protein kinases. Curr Opin Cell Biol 5: 180–186
Sorger P, Murray A (1992) S-phase feedback control in budding yeast independent of tyrosine phosphorylation of $p34^{CDC28}$. Nature 355: 365–368
Sorger P, Severin F, Hyman A (1994) Factors required for the binding of reassembled yeast kinetochores to microtubules in vitro. J Cell Biol 127: 995–1008
Spencer F, Hieter P (1992) Centromere DNA mutations induce a mitotic delay in Saccharomyces cerevisiae. Proc Natl Acad Sci USA 89: 8908–8912
Strunnikov A, Kingsbury J, Koshland D (1995) CEP3 encodes a centromere protein of Saccharomyces cerevisiae. J Cell Biol 128: 749–760
Stueland CS, Lew DJ, Cismowski MJ, Reed SI (1993) Full activation of $p34^{CDC28}$ histone H1 kinase activity is unable to promote entry into mitosis in checkpoint-arrested cells of the yeast Saccharomyces cerevisiae. Mol Cell Biol 13: 3744–3755
Sugino A (1995) Yeast DNA polymerases and their role at the replication fork. Trends Biochem Sci 20: 319–323
Tomkiel J, Cooke C, Saitoh H, Bernat R, Earnshaw W (1994) CENP-C is required for maintaining proper kinetochore size and for a timely transition to anaphase. J Cell Biol 125: 531–545
Toyn J, Toone W, Morgan B, Johnston L (1995) The activation of DNA replication in yeast. Trends Biochem Sci 20: 70–73
Weinert T, Hartwell L (1988) The RAD9 gene controls the cell cycle response to DNA damage in Saccharomyces cerevisiae. Science 241: 317–322
Weinert T, Hartwell L (1990) characterization of RAD9 of Saccharomyces cerevisiae and evidence that its function acts posttranslationaly in cell cycle arrest after DNA damage. Mol Cell Biol 10: 6554–6564
Weinert T, Hartwell L (1993) Cell cycle arrest of cdc mutants and specificity of the RAD9 checkpoint. Genetics 134: 63–80
Weinert T, Kiser G, Hartwell L (1994) Mitotic checkpoint genes in budding yeast and the dependence of mitosis on DNA replication and repair. Genes Dev 8: 652–655
Yeh E, Skibbens R, Cheng J, Salmon E, Bloom K (1995) Spindle dynamics and cell cycle regulation of dynein in the budding yeast Saccharomyces cerevisiae. J Cell Biol 130: 687–700
Yen T, Compton D, Wise D, Zinkowski R, Brinkley B, Earnshaw W, Cleveland D (1991) CENP-E, a novel human centromere-associated protein required for progression from metaphase to anaphase. EMBO J 10: 1245–1254
Zakian V, Runge K, Wang S (1990) How does the end begin? Formation and maintenance of telomeres in ciliates and yeast. Trends Genet 6: 12–16
Zheng P, Fay D, Burton J, Xiao H, Pinkham J, Stern D (1993) SPK1 is an essential S-phase specific gene of S. cerevisiae that encodes a serine/threonine/tyrosine kinase. Mol Cell Biol 13: 5829–5842
Zirkle R (1970) UV-microbeam irradiation of newt-cell cytoplasm: spindle destruction, false anapha se,and delay of true anaphase. Radiat Res 41: 516–537

Genomic Instability and Its Role in Neoplasia

T.D. Tlsty

1 Genomic Instability and Tumor Progression

The study of how genomic integrity is regulated is important not only in the formation and progression of a neoplasia, but also in how a tumor responds to therapy. Genomic instability has been hypothesized to be a driving force behind multistep carcinogenesis (Nowell 1976). A number of genetic changes are required for a normal cell to become tumorigenic (Foulds 1959; Fearon and Vogelstein 1990). If genomic instability increases the rate at which these alterations occur, then the accumulation of changes and subsequent selection for growth and motility advantage may lead to the formation of a neoplasia. Thus, the new variants generated during tumor progression may be fueled by an underlying genomic instability. Once a cell becomes neoplastic, its evolution may continue to malignancy. Further genetic changes are required to confer metastatic properties on the tumor cell. These properties include the ability to invade surrounding tissues, enter the vasculature, extravasate, and colonize a secondary site. Proficiency at each step is necessary for a tumor cell to become fully metastatic. Genetic alterations are the basis for this acquired variation (Rubin 1987; Liotta et al. 1987). The emergence of drug-resistant or radiation-resistant variants is one of the most disappointing aspects of treating a neoplasia. These variants are generated by the same forces that allow the tumor to become established and progress.

Department of Pathology, PO Box 0506, University of California at San Francisco, San Francisco, CA 94143-0506, USA

Neoplastic growth may result from activation of stimulatory mechanisms (proto-oncogenes) and/or from inactivation of inhibitory mechanisms (tumor suppressor genes). A disturbance of homeostasis in a number of cellular systems is neccessary to progress to malignancy. The most commonly studied cellular system is that of growth control. Improper expression of growth factors and growth factor receptors has long been known to contribute to the development of a neoplasm. More recently, it has been found that overexpression of positive regulators of the cell cycle may contribute to tumor progression. For example, experimental evidence has shown that cyclin D1 overexpression leads to hyperplasia and carcinogenesis in a mouse model system (WANG et al. 1994). Amplification of cyclin D1 has been found in primary tumors (MOTOKURA et al. 1991) and tumor cell lines (LUKAS et al. 1994). Mutations in the negative controls on this system are also found. While some cyclin-dependent kinases (CDKs) are overexpressed in tumor lines (TAM et al. 1994), the loss of the cyclin-dependent inhibitors (CDIs), negative regulators which act as tumor suppressors, is also observed in tumor cells (SERRANO et al. 1993). These data suggest that abnormalities in cell cycle control elements can underlie tumorigenesis. Likewise, disturbance in a cell's ability to move from one place to another (used to invade surrounding tissues, enter the vasculature, and extravasate) is a key factor in malignancy.

The processes of generating neoplastic variants (carcinogenesis) and generating drug-resistant variants shares several similarities:

1. Both of these events are known to occur at a rare frequency. Given the multitude of cells that comprise the body and the number of divisions that each cell goes through, it is significant that the frequency of neoplasia is as low as it is. Likewise, out of a population of tumor cells that is exposed to adverse selection pressures, it is only one out of every 10000 or 100000 that can emerge to survive. The generation of drug-resistant variants often occurs at frequencies that range between 10^{-3} and 10^{-6} in tissue culture models.
2. Both processes are known to be multistep in nature. It is well established that the accumulation of mutations in both oncogenes and tumor suppressor genes leads to the formation of a neoplasia. In the generation of drug-resistant variants, such as those due to gene amplification, simple duplications of genetic material is followed by the formation of more gene copies and the rearrangement of DNA sequences. Deletions, translocations, and further amplifications are known to occur as the tumor cell increases its resistance to physical and chemical therapies.
3. Each process seems to depend on the occurrence of an S phase transition. The mutations that occur and contribute to neoplasia have no effect if the cell does not proliferate (i.e., traverse the S phase of the cell cycle). It is in this phase that the mutations are fixed and passed to the daughter material. Experiments analyzing gene amplification and the generation of drug-resistant variants also point to the involvement of the S phase of the cell cycle. If cells are held at confluence during a pretreatment period that is documented to enhance amplification frequency, no increase occurs. Only if cells are cycling does the enhancement of gene amplification frequency occur in mammalian cells.

4. Both tumor progression and the generation of drug-resistant variants are typified by the emergence of a monoclonal population from a heterogeneous population. In tumor cells it is believed that selective pressure provides the screen that allows for the dominance of a monoclonal population. Those cells having the greatest advantage eventually take over or dominate the proliferative fraction of a neoplastic population. Likewise, in drug-resistant variants, one variant emerges from the heterogeneous parental population that is responding to the selection pressure.
5. Both events are known to be accompanied by widespread chromosomal rearrangements. The rearrangements that take place in tumor progression have been documented and have been reported to increase as the population becomes more malignant. Increased chromosomal rearrangements are also seen as the drug-resistant variants emerge. Deletions, translocations, aneuploidy, and inversions have accompanied the amplification events that provide resistance to certain drugs.
6. Finally, amplification of target genes is known to further each process. Amplification of oncogenes is well documented in the progression of neuroblastomas and other neoplastic samples. It is believed that amplification of these oncogenes provides a selective growth advantage to the cells that contain them. Likewise, amplification of the genes that are targeted by various chemotherapies also provides a selective growth advantage and allows for the emergence of the drug-resistant variants.

2 Systems Which Modulate Genomic Instability

In thinking of cellular processes that would affect the endpoint of genomic instability, several candidates come to mind. It has long been appreciated that the repair capacity of a cell has an important effect on the level of genetic instability found in tumor cells. The early description of human cancer syndromes, such as xeroderma pigmentosum, Bloom's syndrome, etc., and the realization that cells from these patients have problems in DNA metabolism and increased chromosomal abnormalities supports the involvement of DNA repair processes in governing the amount of genomic instability a cell experiences.

Systems which provide for the activation or inactivation of chemical and physical carcinogens could also be envisioned to influence the mutation rate in mammalian cells. Enhancement of enzyme systems which detoxify chemicals should decrease the mutation rate while inactivation of these systems should increase the mutation rate. The field of genetic toxicology is examining the contribution of these systems to human disease.

Perhaps less expected to contribute to genomic instability would be mutations in the cellular systems that govern the rearrangements of the immunoglobulin genes. These recombination and mutation events are part of a carefully controlled developmental system that produces heterogeneity in a lymphocytic cell. If this system is improperly regulated, it could conceivably generate rough rearrangements that could contribute to the activation of oncogenes and the inactivation of tumor suppressor genes. Recent work has documented a contribution of this system to the generation of lymphomas (Kirsh et al. 1993).

Logically, one could also postulate that improprieties in chromatin structure could also contribute to genomic abnormalities. If the genetic material is not properly packaged or cannot navigate the structural changes that are repeatedly necessary as a cell replicates and divides, chromosomal breakage and abnormalities would result. At the present time, the documentation of these types of mutations to the generation of genomic instability has not occurred. A subgroup of chromatin alterations that can lead to changes in gene expression has its basis as an epigenetic event. Preferential methylation of specific genes can contribute to the neoplastic process. Recent studies have identified abnormal methylation as the trigger for turning off the expression of a specific kinase inhibitor in bladder cancer and prostate cancer. The small molecular-weight cyclin-dependent kinase inhibitor p16 has been shown to be silenced by methylation in a good proportion of these tumors. This is the first example of methylation changes that are known to contribute to tumor progression and are found in end-stage carcinomas.

In procaryotes and lower eucaryotes there are cellular systems in place which respond to environmental stress and generate a plethora of genetic changes. These inducible systems, which include the SOS response of *Escherichia coli*, generate mutations as a response to DNA damage and are believed to allow for the generation of genetic diversity that is necessary to overcome adverse growth conditions. The activation or misregulation of this type of system could be a key force in the generation of mutational changes which fuel neoplasia and malignancy.

Finally, alterations in cell cycle control can contribute to genomic instability in a variety of different ways. Recent data in this area have documented several of these contributions, and the ways in which cell cycle control is altered is an active area of study. Cell proliferation in the normal cell is regulated by both positive and negative regulatory mechanisms that control growth state transitions and progression through the cell cycle. Early in G_1, the gap phase just prior to DNA replication, the cell is responsive to exogenous signals which reflect environmental conditions. These signals may convey the presence or absence of nutrients as well as contact inhibition. At the restriction (R) point (in mammalian cells) just prior to DNA replication (S phase), the cell commits to cell division, no longer requiring mitogens for cell cycle progression, and becomes unresponsive to external growth signals (Pardee 1974).

From the R point, cell cycle progression is regulated by endogenous signals primarily from a family of CDKs. The temporal activity of CDKs is regulated by cyclin partners whose levels and activity are phase specific (Sherr 1993). As a result, the activity of a given cyclin *cdk* complex varies throughout the cell cycle. Cyclins D, E, and C are active in G_1, with cyclin E being required for traversing the G_1/S boundary. Cyclin A is required for progression through S phase and for transition to mitosis (M phase). Entry into M is signaled by activation of cyclin B, while degradation of cyclin B allows the cell to prepare for S phase (for a review see PINES 1994). Thus, cell growth is regulated by the strictly regimented progression of cell cycle.

3 Genomic Integrity and Cell Cycle Progression

During the proliferation and differentiation of a normal cell there are four stages where the integrity of the genome may be compromised resulting in spontaneous mutations and chromosome alterations. First, errors in DNA sequence are most likely to occur during S phase when the total genome is replicated prior to cell division. Second, during cell division, replicated DNA must be partitioned evenly into two daughter cells. This may provide an opportunity for improper chromosome segregation resulting in abnormal chromosome numbers. Third, repair of damaged DNA can provide an opportunity for errors in DNA sequence to occur. Finally, genomic rearrangements, such as the VDJ rejoining of immunoglobulin chains, can go awry and result in abnormal recombination events.

During the processes of proliferation and differentiation, one might expect to observe many more spontaneous mutations than are experimentally determined (10^{-7}/cell per generation; ELMORE et al. 1983) given the amount of DNA to be duplicated and the number of cell divisions occurring during the life span of a cell. Relatively few errors are detected, however, because the cell is equipped with mechanisms designed to prevent the propagation of errors. Normal cells are able to maintain the fidelity of the genomic sequence through proofreading mechanisms and repair processes.

Progression of the cell cycle relies on a delicate balance of timing. Replication should precede division and should be completed before division begins. Therefore, not only is the order of events important, but an early event must be completed before a later event begins. This order is ensured by the imposition of safeguard mechanisms which evaluate where the cell is in its cycle, mechanisms termed "checkpoints" by HARTWELL and WEINERT (1989). Genes controlling these points serve to generate an inhibitory signal when upstream events are delayed, thus preventing onset of the next downstream event. Yeast and somatic cells have evolved checkpoint controls that ensure that a later event does not occur if the previous event has not yet been completed. Checkpoint controls delay cell cycle progression, presumably to permit repair of errors incurred during DNA replication or chromosome segregation.

Three of the best-defined checkpoints in the cell cycle, the G_1/S, S phase, and the G_2/M transitions, are negatively regulated in response to DNA damage. Cell cycle progress may be delayed to allow repair of damage before entering either S or M phase. At these growth state transitions, DNA damage would be replicated or DNA would be lost due to chromosome breaks. HARTWELL (1992) suggests that checkpoints may possess three types of activities including: "1) surveillance mechanisms that can detect DNA damage, 2) signal transduction pathways that transmit and amplify the signal to the replication or segregation machinery, and 3) possibly repair activities." Mutants that abolish the arrest or delay its occurrence in wild-type cells in response to DNA damage have been isolated (WEINERT and HARTWELL 1988; MORENO and NURSE 1994), indicating that the two checkpoints are under genetic control. These checkpoints are instrumental in facilitating cell viability after damage and in determining whether cells will survive with heritable genetic alterations. All somatic cells display a variety of such checkpoints that monitor cell growth, DNA replication, and DNA repair (HUNTER and PINES 1994).

4 Consequences of Loss of Cell Cycle Control

Cell proliferation is controlled by regulatory mechanisms, both positive and negative, that control growth state transitions and cell cycle progression. Relaxation of these control mechanisms may result in an accumulation of genetic alterations and/or uncontrolled growth. Chromosomal alterations, such as deletions, translocations, amplifications, aneuploidy, and polyploidy, are often observed in tumor cells. These rearrangements may result from defective checkpoints controlling cellular surveillance mechanisms and lead to errors in replication, nondisjunction, and improper chromosome segregation (HARTWELL and KASTAN 1994). In fact, experimental evidence has proven that chromosomal aberrations are evident in cells lacking proper checkpoint control (NISHIMOTO et al. 1978; WEINERT and HARTWELL 1988; WHITE et al. 1994). In G_2, many yeast strains are most resistant to ultraviolet and ionizing radiation due to efficient repair mechanisms (CHANET et al. 1973; BRUNBORG and WILLIAMSON 1978). It is likely that an arrest in G_2 facilitates this repair. In mammalian cells, failure to arrest in G_2 following nitrogen mustard treatment is associated with nuclear fragmentation and enhanced lethality (LAU and PARDEE 1982). Temperature-sensitive *BN-2* mutants in BHK cells and *rad9* mutants in *Saccharomyces cerevisiae* are deficient in G_2 checkpoint control when shifted to nonpermissive temperature or irradiated, respectively, and enter mitosis with unrepaired DNA, resulting in chromosome breaks and cell death (NISHIMOTO et al. 1978; WEINERT and HARTWELL 1988). These checkpoint mutants show elevated rates of mutation and reduced viability (WEINERT and HARTWELL 1990).

Even in human cells, one may observe chromosomal instability in the form of chromosomal rearrangements and gene amplification when cell cycle regulation is compromised (LIVINGSTONE et al. 1992; SCHAEFER et al. 1993; WHITE et al. 1994), suggesting that loss of cell cycle control may be an underlying cause of genomic instability. Recent work (WHITE et al. 1994) demonstrates that compromise of checkpoint function results in the loss of genomic integrity and that integrity can be regained upon restoration of checkpoint function (YIN et al. 1992).

5 Genomic Instability and Gene Amplification

Genomic instability, the increased rate of chromosomal alterations, can take many forms: point mutation, deletion, inversion, translocation, gene amplification, and aneuploidy. Measurements of the rates of spontaneous point mutations have failed to show consistent correlations with tumorigenicity. Some studies support a correlation between spontaneous point mutation rates and tumorigenicity (SESHADRI et al. 1987), and some studies do not (ELMORE et al. 1983). Another marker for instability, gene amplification, has shown a more consistent correlation with tumorigenicity (reviewed in TLSTY et al. 1993) with normal and tumorigenic cells showing vastly different propensities for this phenotype.

Gene amplification was first postulated by BIEDLER and SPENGLER (1976) in methotrexate (MTX)-resistant Chinese hamster cells containing "strikingly long marker chromosomes." Later the mechanism of MTX resistance was determined to be increased copy number (amplification) of *DHFR* genes on these marker chromosomes, resulting in increased expression of DHFR protein (NUNBERG et al. 1978; SCHIMKE et al. 1978; DOLNICK et al. 1979). Subsequently, gene amplification has been determined to be a common mechanism of resistance to a number of agents (reviewed in STARK and WAHL 1984; SCHIMKE 1984).

Amplification of the *CAD* gene is the primary mechanism of resistance to the drug *N*-(phosphonoacetyl)-L-aspartate (PALA). PALA inhibits the aspartate transcarbamylase activity of the trifunctional CAD protein which contains carbamyl phosphate synthetase, aspartate transcarbamylase, and dihydrooratase activities. These enzymes catalyze the first three steps of UMP biosynthesis (WAHL et al. 1979). PALA specifically blocks de novo pyrimidine nucleotide biosynthesis (SWYRYD et al. 1974). Inhibition of aspartate transcarbamylase by PALA therefore results in decreased pyrimidine pools for DNA synthesis. Although amplification of CAD is the major mechanism of resistance, aneuploidy has been detected in virally transformed lines (SCHAEFER et al. 1993; WHITE et al. 1994). Amplification is also seen in tumors where the selection pressure is not apparent (MCGILL et al. 1993; BAND et al. 1989; COX et al. 1965). In many instances, the amplified DNA contains multiple copies of identified oncogenes and correlates with poor prognosis (SEEGER et al. 1985; SLAMON et al. 1987; HENRY et al. 1993). Since amplification is the mechanism of resistance to many agents, including those used in chemotherapy, and is often an indicator of more aggressive tumors, it is important to understand the regulation of gene amplification. For these studies, genomic instability is defined as the ability to generate colonies in PALA due to increased CAD gene copy number. Copy number may be increased by gene amplification or aneuploidy.

A number of models have been proposed as mechanisms of gene amplification (for review see STARK and WAHL 1984; HAMLIN et al. 1991): (a) unequal sister chromatid exchange; (b) rereplication at a locus in a single-cell cycle; (c) episome formation of a locus deletion; and (d) rolling replication. Analysis of early products of *DHFR* gene amplification in CHO cells demonstrates that a double strand break may serve as an initiating event in gene amplification (MA et al. 1993). Chromosomes containing these breaks may fuse at the ends forming dicentrics. These dicentrics are then subject to further cycles of breakage and fusion during chromosome segregation in mitosis.

It appears that chromosome breaks are needed for amplification and other chromosome rearrangements such as inversions and translocations. DNA double-strand breaks initiate amplification and also serve as a signal for cellular DNA damage response (NELSON and KASTAN 1994). The p53 protein induction, a key feature of this response, occurs after double-strand breaks are introduced into DNA by endonucleases (NELSON and KASTAN 1994). Any number of mutations could alter the efficiency with which these breaks are detected by the cell and/or repaired. Questions concerning the relationship between genomic instability, tumor progression, regulation of amplification potential, cell cycle, and cellular response to DNA damage can have important ramifications for therapy.

6 Cellular Response to DNA Damage

The correlation between loss of cell cycle control and amplification potential has led to an analysis of effectors of cell cycle regulation. Inactivation of wild-type p53 function leads to an increased rate of genomic instability (LIVINGSTONE et al. 1992; WHITE et al. 1994; YIN et al. 1992) and tumorigenesis (LI et al. 1988; DONEHOWER et al. 1992). It is thought that the p53 protein acts as a checkpoint control in the cell cycle to permit the repair of damaged DNA by executing cell cycle arrest in G_1 (KASTAN et al. 1991). Loss of the G_1 arrest function in p53 mutant cells results in the accumulation of mutations and genomic instability with the potential to contribute to malignant progression. In cells where p53 is mutated and no G_1 arrest occurs following damage, a G_1 arrest can be restored by the introduction of wild-type p53 (YIN et al. 1992).

In addition to being implicated in amplification and G_1 checkpoint function (WHITE et al. 1994; YIN et al. 1992), p53 has been shown to be instrumental in a cascade which mediates cellular growth arrest in response to DNA-damaging agents (KUERBITZ et al. 1992) including PALA. The transcriptional transactivation activity of p53 suggests that p53 can act indirectly to arrest cell growth. A gene transcriptionally responsive to p53 has been identified, $p21^{WAF1/CIP1}$, and has been demonstrated to inhibit the activity of cyclin *cdk* complexes which regulate cell cycle progression (XIONG et al. 1993; HARPER et al. 1993). Overexpression of p21 protein delays cell cycle progression (GYURIS et al. 1993), and thus p21 appears to be an effector of p53-mediated growth-suppressive activity. The evaluation of the function of the p53/p21 response to DNA damage in normal cells and tumor lines is ongoing and is expected to yield molecular targets for more effective chemotherapy. It is the detection and processing of DNA lesions along with their choreographed sequelae that will provide the variables that determine which cells are vulnerable to probability events that generate genomic rearrangements.

References

Band V, Zajchowski D, Stenman G, Morton CC, Kulesa V, Connolly J, Sager R (1989) A newly established metastatic breast tumor cell line with integrated amplified copies of ERBB2 and double minute chromosomes. Genes Chrom Cancer 1: 48–58

Biedler JL, Spengler BA (1976) Metaphase chromosome anomaly: association with drug resistance and cell-specific products. Science 191: 185–187

Brunborg G, Williamson DH (1978) The relevance of the nuclear division cycle in radiosensitivity in yeast. Mol Gen Genet 162: 277–286

Chanet R, Williamson DH, Moustacchi E (1973) Cyclic variations in killing and 'petite' mutagenesis induced by ultraviolet light in synchronized yeast strains. Biochem Biophys Acta 324: 290–299

Cox D, Yuncken C, Spriggs AI (1965) Minute chromatin bodies in malignant tumors of childhood. Lancet 1: 55–58

Dolnick BJ, Berenson RJ, Bertino JR, Kaufman RJ, Nunberg JH, Schimke RT (1979) Correlation of dihydrofolate reductase elevation with gene amplification in a homogeneously staining chromosomal region in L5178Y cells. J Cell Biol 83: 394–402

Donehower LA, Harvey M, Slagle BL, McArthur MJ, Montgomery CA, Butel JS, Bradley A (1992) Mice deficient for p53 are developmentally normal but susceptible to spontaneous tumors. Nature 356: 215–221

Elmore E, Kakunaga T, Barrett JC (1983) Comparison of spontaneous mutation rates of normal and chemically transformed human skin fibroblasts. Cancer Res 43: 1650–1655

Fearon ER, Vogelstein B (1990) A genetic model for colorectal tumorigenesis. Cell 61: 759–767

Foulds L (1959) Neoplastic development, vol 2. Academic, New York

Gyuris J, Golemis E, Chertkov H, Brent R (1993) Cdi1, a human G1 and S phase protein phosphatase that associates with Cdk2. Cell 75: 791–803

Hamlin JL, Leu T-H, Vaughn JP, Ma C, Dikwel PA (1991) Amplification of DNA sequences in mammalian cells. Prog Nucleic Acid Res 41: 203–239

Harper JW, Adami GR, Wei N, Keyomarsi K, Elledge SJ (1993) The p21 Cdk-interacting protein Cip1 is a potent inhibitor of G1 cyclin-dependent kinases. Cell 75: 805–816

Hartwell L (1992) Defects in a cell cycle checkpoint may be responsible for the genomic instability of cancer cells. Cell 71: 543–546

Hartwell L, Kastan M (1994) Cell cycle control and cancer. Science 266: 1821–1828

Hartwell L, Weinert T (1989) Checkpoints: controls that ensure the order of cell cycle events. Science 246: 629–634

Henry JA, Hennessy C, Levett DL, Lennard TW, Westley BR, May FE (1993) Int-2 amplification in breast cancer: association with decreased survival and relationship to amplification of c-erbB-2 and c-myc. Int J Cancer 53: 774–780

Kastan MB, Onyekwere O, Sidransky D, Vogelstein B, Craig RW (1991) Participation of p53 protein in the cellular response to DNA damage. Cancer Res 51: 6304–6311

Kirsh IR, Abdallah JM, Bertness VL, Hale M, Lipkowitz S, Lista F, Lombardi DP (1994) Lymphocyte-specific genetic instability and cancer. Cold Spring Harb Symp Quant Biol 59: 287–295

Kuerbitz SJ, Plunkett BS, Walsh WV, Kastan MB (1992) Wild-type p53 is a cell cycle checkpoint determinant following irradiation. Proc Natl Acad Sci USA 89: 7491–7495

Lau CC, Pardee AB (1982) Mechanism by which caffeine potentiates lethality of nitrogen mustard. Proc Natl Acad Sci USA 79: 2942–2946

Li F, Fraumeni JF, Mulvihill JJ, Blattner WA, Dreyfus MG, Tucker MA, Miller RA (1988) A cancer family syndrome in twenty-four kindreds. Cancer Res 48: 5358–5362

Liotta LA, Guirguis R, Stracke M (1987) Review article: biology of melanoma invasion and metastasis. Pigment Cell Res 1: 5–15

Livingstone LR, White AE, Sprouse J, Livanos E, Jacks T, Tlsty TD (1992) Altered cell cycle arrest and gene amplification potential accompany loss of wild-type p53. Cell 70: 923–935

Lukas J, Pagano M, Staskova Z, Draetta G, Bartek J (1994) Cyclin D1 protein oscillates and is essential for cell cycle progression in human tumor cell lines. Oncogene 9: 707–718

Ma C, Martin S, Trask B, Hamlin JL (1993) Sister chromatid fusion initiates amplification of the dihydrofolate reductase gene in Chinese hamster cells. Genes Dev 7: 605–620

McGill JR, Beitzel BF, Nielson JL, Walsh JT, Drabek SM, Meador RJ, Von Hoff DD (1993) Double minutes are frequently found in ovarian carcinomas. Canc Genet Cytogenet 71: 125–131

Moreno S, Nurse P (1994) Regulation of progression through the G1 phase of the cell cycle by the rum1 + gene. J Clin Oncol 367: 236–242
Motokura T, Bloom T, Kim HG, Juppner H, Ruderman JV, Kronenberg HM, Arnold A (1991) A novel cyclin encoded by abcl1-linked candidate oncogene. Nature 350: 512–515
Nelson WG, Kastan MB (1994) RDNA Strand Breaks: the DNA template alteration that trigger p53-dependent DNA damage response pathways. Mol Cell Biol 14: 1815–1823
Nishimoto TM, Eilen E, Basilico C (1978) Premature chromosome condensation in a ts DNA-mutant of BHK cells. Cell 15: 475–483
Nowell P (1976) The clonal evolution of tumor cell populations. Science 194: 23–28
Nunberg JH, Kaufman RJ, Schimke RT, Urlaub G, Chasin LA (1978) Amplified dihydrofolate reductase genes are localized to a homogeneous staining region of a single chromosome in a methotrexate-resistant Chinese hamster ovary cell line. Proc Natl Acad Sci USA 75: 5553–5556
Pardee A (1974) A restriction point for control of normal animal cell proliferation. Proc Natl Acad Sci USA 71: 1286–1290
Pines J (1993) Trends Biochem Sci 18: 195–197
Rubin H (1987) The source of heritable variation in cellular growth capacities. Cancer Metastasis Rev 6: 85–89
Schaefer DI, Livanos EM, White AE, Tlsty TD (1993) Multiple mechanisms of N-(phosphonoacetyl)-L-aspartate drug resistance in SV40-infected precrisis human fibroblasts. Cancer Res 53: 4946–4951
Schimke RT (1984) Gene amplification, drug resistance, and cancer. Cancer Res 44: 1735–1742
Schimke RT, Kaufman RJ, Alt FW, Kellems RF (1978) Gene amplification and drug resistance in cultured murine cells. Science 202: 1050–1055
Seeger RC, Brodeur GM, Sather H, Dalton A, Siegel SE, Wong KU, Hammond D (1985) Association of multiple copies of the n-myc oncogene with rapid progression of neuroblastomas. N Engl J Med 313: 1111–1116
Serrano M, Hannon GJ, Beach D (1993) A new regulatory motif in cell-cycle control causing specific inhibition of cyclin D/CDK4. Nature 366: 704–707
Seshadri R, Kutlaca RJ, Trainor K et al (1987) Mutation rate of normal and malignant human lymphocytes. Cancer Res 47: 407–409
Sherr CJ (1993) Mammalian G1 cyclins. Cell 73: 1950–1965
Slamon DJ, Clark GM, Wong SG, Levin WJ, Ullrich A, McGuire WL (1987) Human breast cancer: correlation of relapse and survival with amplification of the HER-2/neu oncogene. Science 235: 177–182
Stark GR, Wahl GM (1984) Gene amplification. Annu Rev Biochem 53: 447–491
Swyryd EA, Seaver SS, Stark GR (1974) N-(phosphonacetyl)-L-aspartate, a potent transition state analog inhibitor of aspartate transcarbamylase, blocks proliferation of mammalian cells in culture. J Biol Chem 249: 6945–6969
Tam SW, Theodoras AM, Shay JW, Draeta GF, Pagano M (1994) Differential expression and regulation of cyclin D1 protein in normal and tumor human cells: association with Cdk4 is required for cyclin D1 function in G1 progression. Oncogene 9: 2663–2674
Tlsty TD, Jonczyk P, White A et al (1993) Loss of chromosomal integrity in neoplasia. Cold Spring Harbor Symp Quant Biol 58: 645–654
Wang TC, Cardiff RD, Zukerberg L, Lees E, Arnold A, Schmidt EV (1994) Mammary hyperplasia and carcinoma in MMTV-cyclin D1 transgenic mice. Nature 369: 669–671
Wahl GM, Padgett RA, Stark GR (1979) Gene amplification causes overproduction of the first three enzymes of UMP synthesis in N-(phosphonoacetyl)-L-aspartate-resistant hamster cells. J Biol Chem 254: 8679–8689
Weinert TA, Hartwell LH (1988) The RAD9 gene controls the cell cycle response to DNA damage in Saccharomyces cerevisiae. Science 241: 317–322
Weinert TA, Hartwell LH (1990) Characterization of RAD9 of Saccharomyces cerevisiae and evidence that its function acts posttranslationally in cell cycle arrest after DNA damage. Mol Cell Biol 10: 6554–6564
White AE, Livanos EM, Tlsty TD (1994) Differential disruption of genomic integrity and cell cycle regulation in normal human fibroblasts by the HPV oncoproteins. Genes Dev 8: 666–677
Xiong Y, Hannon GJ, Zhang H, Casso D, Kobayashi R, Beach D (1993) p21 is a universal inhibitor of cyclin kinases. Nature 366: 701–704
Yin Y, Tainsky MA, Bischoff FZ, Strong LC, Wahl GM (1992) Wild-type p53 restores cell cycle control and inhibits gene amplification in cells with mutant p53 alleles. Cel 70: 937–948

Role of DNA Excision Repair Gene Defects in the Etiology of Cancer

J.M. FORD and P.C. HANAWALT

1 Introduction

Experimental and clinical evidence strongly suggests that exposure to environmental carcinogens is a critical event in the development of the majority of human cancers. Mutations in genes important for normal cellular functions and growth properties, including many proto-oncogenes and tumor suppressor genes, may result from spontaneous or environmentally induced alterations in DNA and contribute directly to the multistage process leading to malignancy. In addition, alterations in the specific genes required for processing and responding to DNA damage may result in an enhanced rate of accumulation of additional mutations, recombinational events, chromosomal abnormalities, and gene amplification (LOEB 1991). Therefore, the removal of lesions from DNA is essential not only for the basic processes of transcription and replication necessary for cellular survival,

Department of Biological Sciences, Stanford University, Stanford, CA 94305, USA but also for maintaining genomic stability and avoiding the development of malignancies.

Many recently converging lines of experimental evidence reveal the complexity of the cellular responses to DNA damage and their role in malignant transformation. A myriad of interrelated biochemical pathways exist which influence: (a) the metabolism of potentially mutagenic or carcinogenic agents; (b) the efficiency and manner by which damaged DNA is recognized and repaired; (c) cell cycle progression and the coordination of DNA replication and cell division relative to the repair of lesions; and (d) the decision point determining survival or the active induction of programmed death of cells carrying different types and amounts of DNA damage. The discovery of several classes of DNA repair genes, which when defective result in a predisposition to the development of certain malignancies, highlights the central role of DNA damage responses in neoplastic transformation. For example, inherited mutations in genes involved in post-replication mismatch repair predispose individuals to the development of colon cancer and several other malignancies associated with the hereditary non-polyposis colon cancer phenotype (FISHEL et al. 1993; LEACH et al. 1993; BRONNER et al. 1994; PAPADOPOULOS et al. 1994). Somatic mutations or allelic loss of the DNA damage-inducible cell cycle checkpoint control protein encoded for by the *p53* tumor suppressor gene is the most commonly described genetic abnormality in the majority of all human cancers (HOLLSTEIN et al. 1991), and germ line mutations in the *p53* gene result in the inherited cancer susceptibility disease Li-Fraumeni syndrome (MALKIN et al. 1990). These two important models of hereditary cancer susceptibility are discussed in detail in several other chapters in this volume.

Because of the importance of DNA repair mechanisms in the cellular response to DNA damage, one would predict that deficiencies in repair enzymes, if compatible with viability, would predispose individuals to malignancies. In fact, a number of rare, inherited disorders have been described which appear to be caused by defects in the repair of DNA lesions (Table 1), and several of these are associated with an increased risk of developing certain cancers (HANAWALT and SARASIN 1986). In this chapter, we will focus on the clinical, biochemical, and molecular characteristics of two autosomal recessive hereditary diseases associated with defects in nucleotide excision repair (NER): xeroderma pigmentosum (XP) and Cockayne's syndrome (CS). Since only XP, but not CS, predisposes individuals to an increased incidence of malignancy, an analysis of the molecular defects underlying these disorders and their effect on mutagenesis and carcinogenesis should help to define the role of enzymes involved in NER for malignant transformation, as well as other cellular processes.

Table 1. Major human hereditary disorders that involve defects in processing DNA damage (adapted from HANAWALT and SARASIN 1986)

Syndrome	Clinical features	Increased incidence of cancers	Hypersensitivities
Xeroderma pigmentosum (classical and variant)	Sunlight hypersensitivity Exposed skin epithelioma Frequent skin cancers Often neurologic abnormalities (none in XP variant) Occular defects	+ + +	UV and chemical carcinogens
Cockayne's syndrome	Cachectic dwarfism Mental retardation (microcephaly) Premature aging Sunlight hypersensitivity	–	UV and chemical carcinogens
Ataxia telangiectasia	Cerebellar ataxia Telangiectasia Neurologic deterioration Partial immunodeficiency High incidence of lymphomas Increased risk of breast cancer in heterozygotes	+ + +	Ionizing radiation bleomycin
Fanconi's anemia	Growth retardation Pancytopenia Bone marrow deficiency Anatomical defects Increased incidence of leukemia	+	Bifunctional alkylators
Bloom's syndrome	Sunlight hypersensitivity Growth retardation High incidence of malignancies	+ + +	UV and bifunctional alkylators

2 Nucleotide Excision Repair

DNA repair may be defined as those cellular responses associated with the restoration of the normal nucleotide sequence following events which damage or alter the genome (FRIEDBERG 1985). A wide variety of endogenous and exogenous agents cause various types of damage to DNA, and multiple enzymatic processes exist to repair these different lesions. The most versatile and ubiquitous mechanism for DNA repair is NER, which functions to remove many types of lesions, including bulky base adducts of chemical carcinogens, interstrand and intrastrand cross-links, and UV-induced cyclobutane pyrimidine dimers (CPD) and 6-4 photoproducts. Such lesions may serve as structural blocks to transcription and replication due to distortion of the helical conformation of DNA, and they may also result in mutations if translesional replication occurs or if they are not repaired correctly. Thus, mechanisms for NER are vital for both genomic stability and cellular survival in humans. While most of the proteins involved in eukaryotic NER have now been identified, the basic biochemical model for NER has been based mainly on studies with bacteria and is assumed to be at least functionally analogous in higher organisms. Studies of NER in mammalian cells have benefitted greatly from analysis of cell lines deficient in particular steps in the repair process.

2.1 Biochemical Mechanism of NER

The general pathway for NER has been understood for 30 years since the discovery of pyrimidine dimer excision (SETLOW and CARRIER 1964; BOYCE and HOWARD-FLANDERS 1964) and repair replication (PETTIJOHN and HANAWALT 1964). The sequential steps are illustrated in Fig. 1 and include: (a) recognition of the damaged site; (b) incision of the damaged DNA strand at or near the site of the defect; (c) removal of a stretch of the affected strand containing the lesion; (d) repair replication to replace the excised region with a corresponding stretch of normal nucleotides using the complementary strand as a template; and (e) ligation to join the

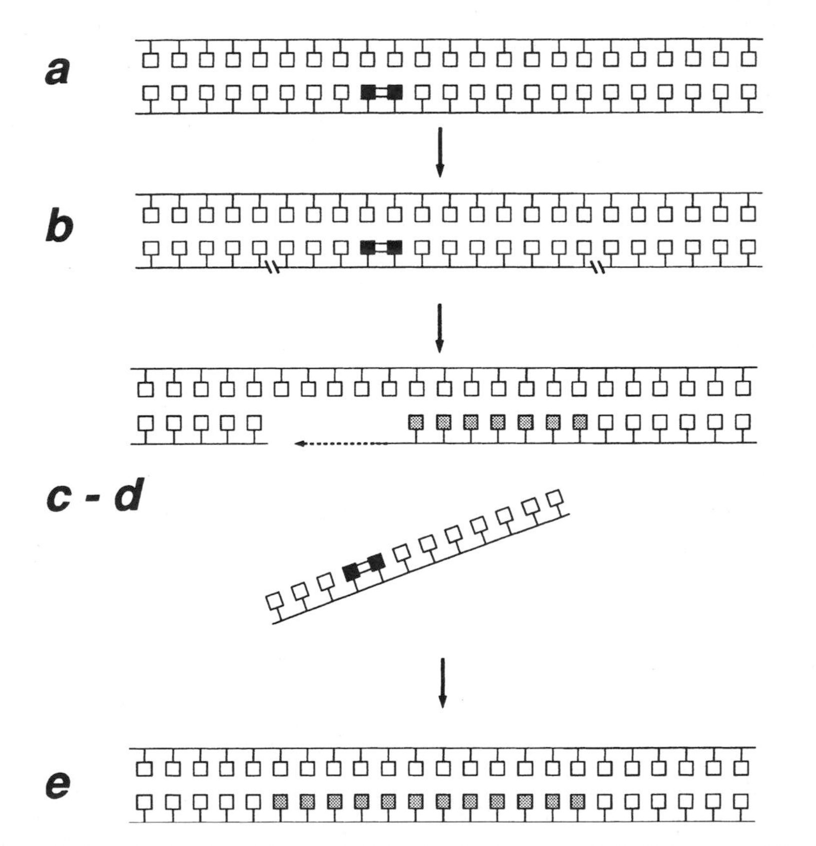

Fig.1 a-e. Nucleotide excision repair. Following the formation of a DNA lesion, the following steps occur sequentially: **a** recognition of the damaged site; **b** incision of the damaged DNA strand near the site of the defect; **c** removal of a stretch of the affected strand containing the lesion; **d** repair replication to replace the excised region with a corresponding stretch of normal nucleotides using the complementary strand as a template; and **e** ligation to join the repair patch to the contiguous parental DNA strand

repair patch at its 3′ end to the contiguous parental DNA strand (Hoeijmakers 1993a,b).

Molecular details of the mechanism of NER are known primarily from work with the prokaryote *Escherichia coli* and have been recently reviewed elsewhere (Hoeijmakers 1993a; Sancar and Tang 1993). Briefly, the basic components of the bacterial system include three proteins, termed UvrA, UvrB, and UvrC, which participate in the initial steps of lesion recognition and incision on both sides of the lesion. Excision of the 12-13-mer damage-containing oligonucleotide requires the concerted action of UvrD (helicase II) and DNA polymerase I. Polymerase I then fills the resultant gap and ligase seals the newly synthesized DNA to the preexisting strand. The cloning and expression of bacterial genes involved in NER has enabled the reconstitution of repair in vitro (Selby et al. 1991; Selby and Sancar 1991, 1993) and has facilitated the identification of additional factors that modulate the activity of NER. As will be discussed more extensively below, investigators in our laboratory first demonstrated in mammalian cells that UV-induced CPD are more efficiently removed from actively expressed genes than from unexpressed regions of the genome (Bohr et al. 1985; Mellon et al. 1986), and that this preferential repair is due to more efficient repair in the transcribed than the non-transcribed DNA strand (Mellon et al. 1987) suggesting a direct association between transcription and repair. This phenomenon was subsequently shown to occur in *E. coli*, in which a pronounced preferential repair of the transcribed strand of the induced *lac* operon was observed compared to that in the non-transcribed strand (Mellon and Hanawalt 1989). Recently, the product of the bacterial *mfd* gene was found to be required for transcription-coupled repair in *E. coli* (Selby et al. 1991), suggesting that it may function as a transcription-repair coupling factor to target NER to lesions within active genes (Selby and Sancar 1993; Hanawalt and Mellon 1993).

To what extent the detailed enzymatic features of prokaryotic NER are preserved in eukaryotes is largely unknown, though the basic steps of lesion recognition, dual incision of the damaged strand and repair synthesis have been clearly documented (Hoeijmakers 1993b). The eukaryotic yeast *Saccharomyces cerevisiae* has been studied as a paradigm for mammalian NER, facilitated by the large number of excision repair-deficient yeast mutants available (Sweder 1994). To date, more than 30 genetic loci have been identified which affect the sensitivity of yeast to UV irradiation, and at least 12 genes have been identified which are specifically involved in NER. While even less is known regarding the specific molecular components of NER in humans, it is certainly likely to be at least as complex as that in yeast. Most of the genes implicated in human excision repair processes have been cloned and demonstrate a remarkable conservation of certain nucleotide sequence motifs with their yeast counterparts.

2.2 Heterogeneity of NER

The preferential repair of expressed genes was first demonstrated when it was discovered in our laboratory that UV-irradiated Chinese hamster ovary cells repair CPD much more efficiently in the expressed dihydrofolate reductase (*DHFR*) gene than in an unexpressed DNA sequence downstream or in the genome overall (Bohr et al. 1985). Similarly, a preferential rate of repair of the *DHFR* gene was observed in human cells (Mellon et al. 1986), though overall genomic repair was significantly more efficient than in murine or rodent cells. We suggested that the blocked RNA polymerase might serve as a signal to recruit repair enzymes to these sites in expressed genes (Mellon et al. 1986). It was subsequently shown that the preferential repair of the expressed *DHFR* gene in hamster and human cells was primarily due to preferential repair of the transcribed DNA strand (Mellon et al. 1987). This so-called transcription-coupled repair has since been demonstrated in a number of other expressed mammalian genes (cf. Leadon and Lawrence 1991; Venema et al. 1992; Ford et al. 1994), in bacteria (Mellon and Hanawalt 1989), and in yeast (Smerdon and Thoma 1990; Sweder and Hanawalt 1992; Leadon and Lawrence 1992) implicating it as a highly conserved and ubiquitous molecular repair pathway.

A number of experimental approaches suggest that the phenomenon of transcription-coupled DNA repair is strictly dependent upon RNA polymerase II transcriptional elongation activity. For example, strand-selective repair is not observed in human and rodent ribosomal RNA genes which are transcribed by polymerase I (Vos and Wauthier 1991; Christians and Hanawalt 1993), and it is abolished by the RNA polymerase II inhibitor, α-amanatin (Christians and Hanawalt 1992). Furthermore, in yeast containing a temperature-sensitive mutation in the *rpb1-1* subunit of RNA polymerase II, the more rapid and preferential repair of the transcribed strand of active genes was lost at the non-permissive temperature, whereas repair on the non-transcribed strand and genome overall was not significantly affected (Sweder and Hanawalt 1992; Leadon and Lawrence 1992). Therefore, it appears that some NER is directly associated with transcription, and a great deal has been recently learned regarding this association (Hanawalt et al. 1994). We have proposed a model for transcription-coupled repair in which a transcription-repair coupling factor specifically binds to the blocked RNA polymerase at a site of transcription blockage and provides both a high-affinity binding site for the repair complex involved in initiating NER and also facilitates the removal of RNA polymerase and truncation of the nascent RNA transcript to provide access for repair enzymes to the lesion site (Hanawalt 1993). In fact, recent studies have shown that transcript shortening occurs at the site of a CPD and is dependent on the transcription factor SII (TFIIS), which may also be essential for the resumption of transcription after the template has been repaired (Donahue et al. 1994).

The existence of a mechanism to facilitate the preferential repair of the transcribed strand of active genes in both eukaryotes and prokaryotes raises a number of questions as to its evolutionary role. Certainly, one could presume that strand-specific repair of active genes would be important for maintaining genomic stability in multicellular organisms by helping to avoid transforming mutations in expressed proto-oncogenes and tumor suppressor genes in humans. However, the lack of an increased incidence of malignancy in individuals with CS, a disease in which transcription-coupled repair has been selectively lost but overall genomic NER retained, argues against this NER pathway as being critical in the process of transformation. This issue will be further explored in a later section with regard to actual mutational spectra in cancers from normal compared to NER-deficient patients. Furthermore, the existence of transcription-coupled repair in unicellular and prokaryotic organisms suggests that its function is more important to the basic processes of transcription and replication required for cellular survival.

2.3 NER Genes

Considerable progress has been made recently in identifying and characterizing human NER genes (Weeda et al. 1993). The most successful strategy has involved identifying human genomic sequences that correct the UV sensitivity and DNA repair deficiency in mutant rodent cell lines. Corresponding human cDNA clones have been isolated and termed excision repair cross-complementing (*ERCC*) genes. Many of the identified genes have been shown to be mutated in human NER deficiencies, such as XP and CS (Table 2). For example, the *ERCC6* gene was first cloned from a human sequence that corrected the UV-sensitive Chinese hamster UV61 cell mutant (Troelstra et al. 1990) and was subsequently found to correct the molecular defect in human CS-B cells, which lack transcription-coupled repair (Troelstra et al. 1992), but not to correct the repair defect in CS-A cells. A human CS-A cDNA was subsequently cloned by functional complementation of CS-A cells, and lacked the ability to restore UV-resistance to CS-B cells (Henning et al. 1995). Thus, the CS-A and CS-B proteins, perhaps as a heterodimer, appear to be directly involved in linking DNA repair and transcription, and may function in an analogous manner to the prokaryotic *mfd* transcription-repair coupling factor.

Table 2. Cloned human nucleotide excision repair genes

Human gene	Mutant rodent cell line(s)	Human disease	UV sensitivity
ERCC1	UV20, 43-3B	?	+++
ERCC2	UV5, V-HI	XP-D	+++
ERCC3	UV24	XP-B/CS-C	+++
ERCC4	UV41, UV47	XP-F	+++
ERCC5	UV135	XP-G	++
ERCC6	UV61	CS-B	+
XPA		XP-A	+++
XPC		XP-C	+
CSA		*CSA*	+

Another recently identified human repair gene, *ERCC3*, has been implicated in patients with XP-B/CS-C, a rare group of XP patients who also display clinical features of CS and are completely deficient in the repair of overall genomic DNA, including transcribed sequences. *ERCC3* may thus constitute another functional element involved in transcription-repair coupling (WEEDA et al. 1990). However, this gene is also essential for overall genomic excision repair (SWEDER and HANAWALT 1993, 1994; HANAWALT et al. 1994; MA et al. 1994). Similarly, the human repair gene ERCC2 has been cloned and found mutant in patients with XP-D/CS (TAKAYAMA et al. 1995). Remarkably, both the ERCC2 and ERCC3 gene products have now been shown to also be components of the basal transcription initiation factor TFIIH (SCHAEFFER et al. 1993; VAN VUUREN et al. 1993; DRAPKIN et al. 1994; SCHAEFFER et al. 1994), and to possess helicase activities which may help in open complex formation and promoter clearance. Indeed, the convergence at the molecular level of NER and transcription is supported by a growing number of reports in which essential protein components of the eukaryotic transcription factor TFIIH are required for NER (reviewed in HANAWALT et al. 1994; FRIEDBERG et al. 1994). The presumed dual role for many of these proteins in both NER and gene transcription therefore provides new and provocative hypotheses regarding the cellular and clinical consequences of mutations in these genes. For example, it has been suggested that some of the developmental abnormalities in CS may in fact represent a class of "transcription disease," rather than consequences of a repair deficiency (BOOTSMA and HOEIJMAKERS 1993).

3 Human NER-Deficient Syndromes and Cancer

A direct correlation between unrepaired DNA damage and carcinogenesis in humans was first established when CLEAVER (1968) found that the cancer-prone hereditary disease XP involved a defect in the repair of DNA lesions produced by UV light. Since that discovery, a great deal has been learned regarding the clinical characteristics and molecular defects of this and the related UV hypersensitivity disorder CS. In this section, we will review these findings and discuss the potential role of specific defects in NER in oncogenesis.

3.1 Clinical Characteristics

Xeroderma pigmentosum is a rare, autosomal recessive disease in which homozygous individuals display several characteristics: (a) extreme sensitivity of the skin to sun (UV) exposure evident by 1 year of age; (b) pigmentation abnormalities and premalignant lesions in sun-exposed skin; (c) up to 4000-fold increases in incidence of skin cancers (predominantly squamous and basal cell carcinomas, but also melanomas) and ocular neoplasms, occurring 3–5 decades earlier than in the general population; and (d) a 10- to 20-fold increased incidence of internal cancers in non-sun-exposed sites (Kraemer et al. 1984; Cleaver and Kraemer 1989). Overall, there is an approximately 30-year reduction in life span among XP patients, and many die due to malignancies (Kraemer and Slor 1985). Approximately 20% of XP patients also display progressive neurologic degeneration, characterized by peripheral neuropathy, sensorineural deafness, progressive mental retardation, and cerebellar and pyramidal tract involvement (Robbins 1988). The incidence of XP is approximately one in 250 000 individuals in the United States and in Europe, but significantly higher (one in 40 000) in Japan and Egypt.

Cockayne's syndrome is also an autosomal recessive disease with defective repair of UV-damaged DNA (Schmickel et al. 1977) and is characterized by cutaneous photosensitivity, cachectic dwarfism, skeletal abnormalities, retinal degeneration, cataracts, severe mental retardation, and neurologic degeneration characterized by primary demyelination, calcific deposits in the brain, and normal pressure hydrocephalus (Robbins 1988; Timme and Moses 1988). However, in contrast to patients with XP, those with CS are not at an increased risk for developing skin cancers. The average life span of individuals with CS is only 12 years, with most patients succumbing to infectious or renal complications, rather than cancer (Nance and Berry 1992).

3.2 Complementation Groups

Cultured fibroblasts from XP patients are hypersensitive to UV irradiation and to carcinogens that form bulky adducts in DNA. The biochemical defect in cells from most XP individuals is in NER (Cleaver 1968), though in a small group of cases (termed XP variants), excision repair appears normal, and a defect apparently exists in bypass replication at unrepaired lesions (Wang et al. 1993). Complementation analysis via fusion of cells or cell nuclei from different patients has demonstrated genetic heterogeneity within XP and provided evidence for the existence of at least seven excision-deficient complementation groups, termed XP-A through XP-G, in addition to the above-mentioned XP variant (Table 3). Similarly, CS is characterized by the existence of at least three complementation groups, termed CS-A through CS-C. Presumably, each XP and CS complementation group represents a defect in a distinct DNA repair gene. Furthermore, three patients have been described in XP group B (Robbins et al. 1974) and several in XP groups D and G who share the DNA repair defects and clinical features of CS together with the cutaneous manifestations of XP (XP-B/CS-C, XP-D/CS, and XP-G/CS, respectively). The two most recently identified patients (siblings) with XP-B are hypersensitive to UV, severely defective in NER, display some cutaneous and neurologic abnormalities common to XP and CS, but have not developed skin cancers even beyond age 40 (Vermeulen et al. 1994). Molecular analysis of the XP-B/CS-C-correcting *ERCC3* gene has revealed a single-base missense mutation in a conserved region for both these patients. Thus, the relationship between defects in repair proteins and UV-induced skin carcinogenesis may, at least for cer-

Table 3. Properties of XP and CS complementation groups

Complementation group	Associated CS	Frequency	UDS (%)	Overall repair	Transcription-coupled repair	Skin cancer	Neurologic abnormalities
XP-A	–	High	< 5	–	–	+	+
XP-B	+	Only 3	< 8	–	–	+/–	+
XP-C	–	High	15–30	–	+	+	–
XP-D	+/–	Intermediate	20–50	–	–	+	+
XP-E	–	Rare	< 50	–	?	–	–
XP-F	–	Rare	15–30	+/–	–	–	–
XP-G	+/?	Rare	5–15	–	?	+/–	+
XP variant	–	High	100	+	+	+	–
CS-A	+	Rare	100	+	–	–	+
CS-B	+	High	100	+	–	–	+

tain genes, be more complex than can be simply accounted for by abnormalities in DNA repair.

Cells from XP complementation groups A through G all appear to be defective in an early step in NER, prior to and including incision. However, they display variable degrees of UV sensitivity and corresponding deficiencies in their capacity to carry out the incision step of NER (Table 3). For example, XP-C cells are only moderately UV sensitive and have been shown to repair limited domains of their genome (Mansbridge and Hanawalt 1983), while the repair deficiency and UV sensitivity is more pronounced for XP groups A, B, and G.

3.3 Biochemical and Molecular Aspects of NER-Deficient Diseases

The development of methods to measure the efficiency and extent of NER within specific regions of the genome (Bohr et al. 1985) has greatly advanced our understanding of the underlying defects within many of the XP and CS complementation groups (Table 3). For example, XP-A cells demonstrate severe loss of both overall genomic and transcription-coupled DNA repair (Evans et al. 1993a) and display the greatest UV sensitivity. XP-C cells have been shown to selectively retain the ability to repair CPD in the transcribed strands of expressed genes (Venema et al. 1990b, 1991), though they are deficient in overall genomic NER (Mansbridge and Hanawalt 1983; Kantor et al. 1990). Therefore, the relative resistance of XP-C cells to UV irradiation, compared to XP-A, may be due to persistent transcription-coupled repair of active genes. This view is supported by studies of a partial revertant of an XP-A cell line called XP129 (Lommel and Hanawalt 1993). The revertant has regained the full UV resistance of normal human fibroblasts and it also has regained transcription-coupled repair. The features of repair in the XP129 cells interestingly correspond to those of most rodent cells. That is, repair of 6–4 photoproducts is efficient throughout the genome, while CPD repair is restricted to transcribed DNA strands. It is concluded that transcription-coupled repair is more important to cell survival than is overall genomic repair, at least for CPD.

In contrast, while CS cells appear normal in their ability to repair overall genomic DNA, MAYNE and LEHMANN (1982) first observed that these cells were defective in the recovery of RNA synthesis following UV irradiation, suggesting a deficiency in the removal of CPD from active genes. Indeed, we (HANAWALT 1993), and others (VENEMA et al. 1990a), have directly demonstrated that CS cells are selectively defective in the preferential repair of transcribed strands of active genes, which likely accounts for the UV sensitivity of these cells. In fact, our studies demonstrated a reduction of repair within the expressed *c-abl* gene to levels even below that of overall genomic repair (HANAWALT and MELLON 1993), suggesting that in the absence of a mechanism for transcription-coupled repair, a stalled RNA polymerase blocks access of recognition and repair proteins to lesions within transcribed DNA strands. However, it has also been reported that CS cells repair the transcribed strand of the active *DHFR* and *ADA* gene at similar rates as the non-transcribed strand (VAN HOFFEN et al. 1993), yet more slowly than either strand in normal cells. The defect in CS may also affect more than simply the preferential repair of polymerase II-transcribed genes, since CHRISTIANS and HANAWALT (1994), recently showed markedly deficient repair of CPD in polymerase I-transcribed ribosomal RNA gene sequences in cell lines from both CS-A and CS-B patients, genes that do not exhibit transcription-coupled repair in normal cells.

Therefore, analysis of cell lines and individuals from several of the XP and CS complementation groups provides a useful model system by which to determine the individual contribution and role of overall and transcription-coupled NER pathways to UV-induced carcinogenesis.

3.4 Relationship Between XP and CS Repair Defects and Clinical Phenotype

Analysis of the specific abnormalities in NER displayed by the various genetic complementation groups of XP and CS allows correlations to be drawn with their heterogeneous clinical features. Specifically, only those subgroups of patients who display a defect in overall repair of genomic DNA are at significantly increased risk for developing UV-induced malignancies. In contrast, the neurologic symptoms associated with certain other complementation groups of XP and CS are found only in those groups which are defective in transcription-coupled repair (Table 4). In fact, recent data implicating particular human NER enzymes as also being components of a transcription factor complex suggest that the clinical phenotype of patients with defects in transcription-coupled repair may actually be due to abnormalities in transcription rather than in repair itself, as noted above.

For example, patients with XP-A and XP-D generally possess both severe neurologic disorders and a high incidence of skin cancers, and their cells lack the ability to perform both overall and transcription-coupled repair (EVANS et al. 1993a). Patients with XP-C develop skin disorders, but remain neurologically intact and retain the selective repair of the transcribed strand of active genes. Patients in the XP variant group have mild to severe skin symptoms and are neurologically normal (Cleaver and Kraemer 1989). XP variant cells have intact transcription-coupled repair and overall genomic repair, but are defective in a post-replication repair process. Patients from all CS complementation groups are distinguished by a complex phenotype, including developmental and neurologic abnormalities, but they are not at increased risk for malignancies. As noted above, cells from CS patients are defective in transcription-coupled repair, but are generally intact for overall repair functions.

While potentially explaining the molecular basis for many of the clinical characteristics of XP and CS, these observations present an apparent paradox with regard to these patients' cancer risk. The current theory of multistage carcinogenesis suggests that activating mutations in proto-oncogenes and inactivating mutations in tumor suppressor genes drive the progression of cells from normal to transformed to frankly malignant. Many currently recognized oncogenes and tumor suppressor genes are known to possess important cellular functions, and to be actively expressed in normal cells. Since CS cells are defective in the repair of actively expressed genes and have also been shown to be hypermutable (Mayne et al. 1982), a reasonable prediction would be that these patients would acquire mutations in genes leading to transformation more readily than normal patients. However, this is not supported by the clinical picture.

In contrast, XP patients all share a defect in overall NER and a dramatic propensity to develop UV-induced skin cancers, even though some XP patients retain efficient repair of expressed genes. It is not readily apparent why a defect in overall NER, but not transcription-coupled repair, predisposes individuals to cancer. However, it appears that the more rapid and preferential repair of transcriptionally active genomic regions is mainly important to facilitate the formation of full-length mRNA transcripts and directly affects cellular viability, while repair of transcriptionally silent DNA is much more critical for reducing the mutational load of cells and protecting against neoplastic transformation.

Recent analyses of the DNA damage-inducible, p53-dependent apoptotic pathway in XP and CS cells provides an interesting alternative explanation for this paradox. Yamaizumi and Sugano (1994) first demonstrated that DNA damage-induced activation of the wild-type p53 gene product occurred following lower doses of UV irradiation in human fibroblasts deficient in transcription-coupled repair (XP-A and CS-B) than in those proficient in transcription-coupled repair (XP-C and normal fibroblasts), regardless of the efficiency of global nucleotide excision repair. Ljungman and Zhang (1996) have confirmed and extended these findings to demonstrate that apoptosis also occurs following lower doses of UV irradiation in XP-A and CS-B cells than in normal or XP-C human skin fibroblasts, and that these results correlated with the induction of p53.

These results suggest that UV-induced activation of p53 is evoked specifically by lesions in the transcribed strand of expressed genes, and that perhaps the blockage of RNA polymerase II by UV photoproducts may serve as a signal for the induction of p53 and subsequent apoptosis. Furthermore, these results provide support for a provocative additional model to explain the differences in clinical phenotypes and cancer predisposition among individuals with different nucleotide excision repair deficiency syndromes. In this model, cells from patients defective in transcription-coupled repair are more prone to undergo apoptosis following exogenous or endogenous DNA damage, perhaps leading to the development of the neurological abnormalities seen in these groups. Conversely, enhanced damage-induced apoptotic cell death may result in the elimination of DNA damaged, premutagenic cells, and lower the risk of carcinogenesis (LJUNGMAN and ZHANG 1996).

Another puzzling aspect of the clinical phenotype of XP is why these patients do not appear to be at a greater risk for developing neoplasms other than skin cancers. While a disproportionate number of relatively rare tumors such as brain sarcomas and extra-glossal carcinomas of the oral cavity have been described in XP patients under 40 years of age (KRAEMER et al. 1984), individuals with XP do not appear to be at significantly increased risk for more common solid or hematologic malignancies. Owing to the presumed importance of NER pathways for removing bulky DNA adducts resulting from environmental carcinogen exposure, it is surprising that a greater number of cancers such as lung, liver, breast, colon, gastric, and other gastrointestinal malignancies are not seen in XP patients. Based on this apparent lack of an increased number of internal malignancies, CAIRNS (1981) has argued that such common cancers in the normal population are not caused by the particular DNA lesions which XP patients are unable to repair. However, alternative explanations include a potential sample bias due to the early mortality experienced by XP patients, or a decreased exposure to non-UV environmental carcinogens during early life for XP patients.

3.5 Disease Due to Defects in Transcription

As discussed previously, many recently identified human excision repair proteins have been found to be integral components of the basal transcription factor complex TFIIH. These observations raise the provocative possibility that many of the heterogeneous and systemic clinical symptoms manifested by patients with XP and CS may in fact be due to subtle alterations in gene transcription and regulation, rather than caused directly by the corollary defects in NER. This hypothesis is particularly attractive for explaining the pathogenesis of the seemingly developmentally related abnormalities in these patients, such as neurologic and growth defects. However, it is more difficult to generalize this explanation to include CS-B, since the *ERCC6* gene, which when mutant results in this disease, has not been found to associate with or function as a transcription factor. Furthermore, while a defining feature of CS is a lack of resumption of RNA synthesis following UV irradiation of cells in vitro (MAYNE and LEHMANN 1982), evidence for specific transcriptional abnormalities in the absence of DNA damage has not been reported for CS cells, nor for XP cells which display defects in transcription-coupled repair due to mutations in genes whose products are associated with TFIIH (XP-B/CS-C, XP-D/CS).

4 Role of NER in Mutagenesis

One approach to evaluate the influence and significance of certain DNA repair functions on carcinogenesis and transformation is to analyze patterns of mutagenesis in clinical tumor samples from repair-deficient patients and in experimental systems using cells derived from these patients. The following sections will discuss several experimental approaches employed to probe these questions, by looking at UV-induced mutational frequency and patterns in NER-deficient cells and in clinical samples from patients with deficiencies in NER.

Heterogeneity of DNA repair processing may have implications for induction of mutations since the frequency of mutations will depend in part on the extent of repair of pre-mutagenic lesions prior to replication, as well as the consequences on cellular survival of the damage event or its mutagenic result. Furthermore, preferential repair of transcribed DNA strands should be expected to have measurable consequences for the resulting pattern of mutations. Specifically, transcription-coupled repair should result in an accumulation of DNA adducts in non-active regions of the chromosome and in the non-transcribed strands of active genes, leading to a relative abundance of mutations at these sites. Therefore, the mutational spectrum resulting from UV irradiation will be influenced by the more efficient repair of the transcribed strand and predicts a bias toward mutations resulting from lesions in the non-transcribed strand.

4.1 UV-Induced Mutational Spectra in Shuttle Vectors

The development of a host cell reactivation assay using a replicating plasmid has allowed for the analysis of UV-induced base substitution mutagenesis in various human cells. A "shuttle vector," derived from viral genomes such as SV40 to ensure replication in mammalian cells, and, containing a marker gene as a mutational target, is UV irradiated and transfected into mammalian cells. Following DNA repair, plasmids which have replicated are recovered, screened for mutations, and sequenced (Brash 1988). In this manner, the effect of specific host cell DNA repair enzymes on UV-induced base mutations may be determined for a defined DNA sequence.

Several groups have now used the pZ189 shuttle vector to perform such experiments in various normal, XP, and CS cell lines. This plasmid contains a 150-bp bacterial suppressor tRNA gene, *supF*, that allows for selection of mutants in an indicator strain of bacteria (Bredberg et al. 1986). The mutational spectrum of UV-irradiated pZ189 following replication in cells from normal, repair-proficient individuals demonstrates predominantly C to T transition mutations at dipyrimidine sequences, similar to results previously obtained in bacteria (Brash 1988). This observation has been explained by Tessman (1976) as the "A rule," whereby polymerase preferentially inserts an adenine opposite the site of uninformative damaged bases. Similar experiments utilizing pZ189 in fibroblasts or lymphoblastoid cells from XP-A, XP-C, and XP-D patients consistently demonstrated a higher frequency of mutated plasmids compared to experiments using normal cells, but an altered spectrum of mutations characterized by a relative increase in the frequency of C to T transitions (90% compared to 70%) and a decrease in other mutations, such as transversions and multiple-base substitutions (BREDBERG et al. 1986; SEETHARAM et al. 1987, 1991; YAGI et al. 1991, 1992). These results suggest that UV-induced C to T transition mutations may be particularly important for UV-related skin cancers, especially in patients with XP. In fact, this prediction has been borne out, as discussed below.

The use of this shuttle vector system to probe the effect of transcription-coupled repair on UV-induced mutagenesis is problematic, as it is unlikely that the bacterial *supF* gene is transcribed by RNA polymerase II in human cells. Nevertheless, several investigators have determined the UV-induced mutational spectrum of pZ189 following replication in CS cells and found a pattern similar to that seen with fibroblasts from normal humans (MURIEL et al. 1991; PARRIS and KRAEMER 1993), as would be expected for a non-transcribed DNA sequence. The ability to determine the effect of transcription-coupled repair on UV-induced mutations in a shuttle vector system awaits the construction of an appropriate plasmid containing an inducible, polymerase II-transcribed gene selectable for mutations. Such a vector will prove very interesting for use in XP-C and CS cell lines to determine the effect of the presence and absence, respectively, of transcription-coupled repair on mutational frequency and potential strand bias.

4.2 Experimental UV-Induced Mutations in Endogenous Genes

The use of the polymerase chain reaction has facilitated the study of the effect of NER on mutations in endogenous genes in mammalian cells and tissues. VRIELING et al. (1991) first demonstrated that strand-specific repair of an expressed mammalian gene was associated with a bias toward the non-transcribed strand for the dipyrimidine sites at which the majority of UV-induced mutations occurred by mapping mutations in the *HPRT* gene in several repair-proficient hamster cell lines (which display a CPD-repair phenotype similar to that of human XP-C cells). These results were extended to human cells by Maher and colleagues (McGREGOR et al. (1991), who demonstrated that when synchronized, repair-proficient primary human fibroblasts were UV irradiated in early G_1 of the cell cycle (6 h prior to S phase), 80% of the resulting *HPRT* mutations arose at dipyrimidine sites situated in the non-transcribed strand, suggesting that the bias is due to preferential repair of the transcribed strand prior to replication. In contrast, when the same cells were irradiated in S phase, thus eliminating time for DNA repair prior to replication, or when human XP-A cells were used to eliminate NER entirely, an average of 75% of *HPRT* mutations were due to lesions in the transcribed strand. Similarly, VRIELING et al. (1989) found that 90% of mutations in repair-deficient hamster cells were due to photoproducts in the transcribed strand of *HPRT*. These authors argue that the transcribed strand bias seen for mutations in this gene in the absence of transcription-coupled repair which cannot be explained by the distribution of dipyrimidine sites may be due to differences in polymerase error rate between synthesis of the leading and lagging DNA strand on damaged templates (Mullenders et al. 1991). Alternatively, in the absence of transcription-coupled repair, the blocked RNA polymerase may preclude repair at these sites, thus giving the reverse bias.

Mutations at dipyrimidine sites in the p53 tumor suppressor gene characteristic of UV-induced DNA damage have been identified in a high frequency of human skin cancers, providing genetic evidence for a direct relationship between sunlight and the development of skin cancer (Brash et al. 1991). Murine skin cancers induced by repeated UV-exposure provide an excellent model for investigating the mutagenic effects of UV, by determining the resulting sequences of the *p53* gene. Furthermore, unlike the *HPRT* mutation assay, which relies on phenotypic selection of mutants, this cancer-related gene mutation assay is based on genotype analysis and reflects the carcinogenic as well as mutagenic effects of UV-induced DNA damage. Since the *p53* gene can be functionally mutated at more than 100 sites (in contrast, for example, to the *ras* proto-oncogene), it is a particularly good target for determining in vivo mutational spectra, though most analyses only look at restricted regions of the gene. Kress et al. (1992) reported that 32% of 22 squamous cell skin carcinomas from several strains of mice chronically exposed to UV-B irradiation contained base substitutions in the p53 gene, and all were located at dipyrimidine sites in the non-transcribed strand of the gene. Similarly, analysis of 11 tumor cell lines derived from UV-induced murine skin cancers revealed all contained one or more p53 mutations at dipyrimidine sites, 76% of which arose from lesions in the non-transcribed strand (Kanjilal et al. 1993). Since rodent cells are very poor at repairing CPD in non-transcribed DNA sequences but repair the transcribed strands of active genes quite well (Mellon et al. 1987), these results strongly implicate transcription-coupled repair in causing the resulting mutational strand bias in the p53 gene.

The mutational spectrum of the p53 gene has also been analyzed in UV-irradiated cultured human keratinocytes (Nakazawa et al. 1994). CC to TT transition mutations at two specific codons were measured and found to increase in a UV dose-dependent manner, while no mutations occurred in non-irradiated controls. Similar experiments using cells from XP and CS individuals would be revealing regarding the relative frequency of mutation induction and potential strand bias.

Thus, these data from a variety of experimental systems allow one to make predictions of the mutational spectra and strand bias for UV-associated mutations in genes from tumors arising spontaneously in normal patients compared to those from patients with various NER deficiencies.

4.3 *p53* Mutational Spectra in Human Tumors

Until recently, the role of sunlight exposure in the pathogenesis of skin cancers was inferred based upon strong epidemiological evidence since the molecular basis for these tumors was unclear. However, recent observations show that mutations of the *p53* tumor suppressor gene, the most commonly altered gene in the majority of all human cancers (HOLLSTEIN et al. 1991), occur in over half of human squamous and basal cell carcinomas of the skin, and that the spectrum of these mutations directly implicates UV-induced DNA damage as the primary carcinogenic event (BRASH et al. 1991; ZIEGLER et al. 1993). For example, BRASH et al. and ZIEGLER et al. found that nearly all p53 mutations in human squamous and basal cell carcinomas occurred at dipyrimidine sites, and that a quarter of these resulted from CC to TT double-base transitions, virtually pathogneumonic for UV-induced mutations. This observation has been extended to demonstrate that UV-specific p53 gene mutations occur in normal skin and may be useful as a biologically relevant measure of UV exposure and a possible predictor for skin cancer risk (NAKAZAWA et al. 1994). In this study, 17 of 23 (74%) normal skin samples from sun-exposed sites contained CC to TT transition mutations in the p53 gene, while only one of 20 (5%) samples from non-sun-exposed sites displayed such an alteration in this gene.

An analysis of the strand distribution of p53 transition mutations at dipyrimidine sites in skin cancers provides the first opportunity to assess directly the role of overall and transcription-coupled NER in carcinogenesis of a specific human tumor. In fact, for basal cell carcinomas, a trend has been observed for a higher frequency of mutations at dipyrimidine sites in the non-transcribed strand of p53 in several small series (ZIEGLER et al. 1993; RADY et al. 1992; MOLES et al. 1993), though p53 mutations were evenly distributed between DNA strands in reported squamous cell skin carcinomas (BRASH et al. 1991; PIERCEALL et al. 1991). Of note, a pronounced predominance of guanine transversion mutations arose from lesions in the non-transcribed strand of p53 in human lung cancers (HOLLSTEIN et al. 1991; PUISIEUX et al. 1991; MILLER et al. 1992). The preferential repair of tobacco carcinogen benzo[*a*]pyrene diol epoxide or similar adducts in the transcribed strand of active genes (CHEN et al. 1992) may account for this bias.

Several recent reports have also described the *p53* mutational spectrum in skin cancers from patients with XP. While still insufficient to correlate particular mutational patterns to specific XP complementation groups, the findings as a whole are remarkably similar to those predicted from the in vitro and in vivo model systems previously described. SATO et al. (1993) reported that five of eight nonmelanoma skin cancers from six Arabic XP patients contained mutations in the p53 gene. All base changes were located at dipyrimidine sites containing cytosine residues. Of the four patients for whom complementation groups were known, mutations likely arose at dipyrimidines in the non-transcribed strand of *p53* in an individual with XP-A and one with XP-C, and from the transcribed strand in an individual with XP-F and one XP variant.

DUMAZ et al. (1993) analyzed *p53* mutations in a larger series of XP skin cancers: 48% of 23 squamous cell carcinomas and 27% of 11 basal cell carcinomas of the skin contained *p53* mutations, all at dipyrimidine sites and 61% CC to TT tandem base transitions. Furthermore, 15 of the 17 documented mutations arose from lesions in the non-transcribed strand. Because complemention group information could only be inferred from clinical features and UDS measurements for most patients, a specific correlation between the non-transcribed strand bias for mutations and NER defects cannot be made. However, the authors claimed that at least 50% of the tumors were likely from patients with XP-C, 25% from patients with XP-A, and 25% from XP-variant patients. Of note, one of the two samples in which a mutation arose from a lesion in the transcribed strand was from a patient documented to have XP variant with normal overall NER.

These investigators have gone on to review all reported *p53* mutations in XP and non-XP human skin cancers and compared them to the *p53* mutational spectrum of internal tumors, utilizing a database of nearly 2000 p53 sequences (Dumaz et al. 1994). This analysis is significant for demonstrating that the mutational spectra of *p53* from skin tumors is statistically different from all other reported human cancers, but very similar to those observed in other UV-treated genes in model systems, and that these characteristics were exacerbated in a predictable manner in XP skin tumors, consistent with the absence of efficient NER. For example, 74% of all skin tumor mutations were C to T transitions, whereas only 47% of internal tumors displayed this base substitution pattern, and most of those were located at CpG sequences implicating deamination of 5-methyl cytosine as the mutagenic event. In XP skin tumors, 100% of mutations were located at dipyrimidine sites and 55% of these were tandem CC to TT transitions, compared to only 14% CC to TT double-base mutations in skin cancers from normal individuals. Internal tumors only rarely displayed these double cytosine base mutations (0.1%). Finally, 95% of mutations from XP skin cancers arose from lesions presumed to be on the non-transcribed strand, while internal tumors or non-XP skin cancers did not show this strand bias. This last observation is suggestive of a role for transcription-coupled repair, in the absence of overall NER, for causing a non-transcribed strand bias for *p53* mutations by selectively repairing lesions in the transcribed strand, as would be predicted for patients with XP-C. However, this mechanistic explanation should not hold true for patients from other XP complementation groups in which both overall and transcription-coupled repair are defective (e.g., XP-A). Since the actual number of patients from various complementation groups in these studies remains unknown, it is difficult to assess the significance of the *p53* mutational strand bias and its potential association with transcription-coupled repair.

Taken together, the data from UV-induced mutations in human skin cancers is consistent with, though not proof of, the notion that in the normal population, under conditions leading to skin carcinogenesis, transcription-coupled repair after UV irradiation does not play a fundamental role in avoidance of malignancies. This is in agreement with the previously discussed observations that CS patients who lack transcription-coupled repair are not at an increased risk for UV-induced skin tumors. The mutational frequency and spectra of the *p53* gene in sun-exposed or UV-irradiated normal skin samples from CS individuals would therefore be of interest and further advance this hypothesis.

We (FORD et al. 1994), and others (EVANS et al. 1993b), have now provided direct experimental evidence for preferential repair of the human *p53* gene by measuring the rate of removal of CPD from each strand of this gene in several human fibroblast cell lines. For example, we found more rapid repair of UV-induced DNA damage in the transcribed than in the non-transcribed strand of the *p53* gene in a normal, diploid fibroblast cell line, with repair efficiency of the non-transcribed strand similar to that of other regions of unexpressed DNA (88% repair of the transcribed strand and 60% repair of the non-transcribed strand 24 h following irradiation; FORD et al. 1994). EVANS et al. (1993b) reported similar results for another normal fibroblast cell line and further demonstrated that, in a human XP-C cell line, the transcribed strand of *p53* was repaired with an efficiency similar to that of normal cells, but that the non-transcribed strand displayed no repair at all 24 h after UV irradiation, as expected. Therefore, these results suggest that the repair characteristics of the non-transcribed strand, and by implication those DNA sequences targeted by overall genomic NER mechanisms in general, are most relevant for UV-induced carcinogenesis.

5 Conclusions

Cellular responses to DNA damage are highly complex, with many factors actively influencing the outcome of damage events, whether resulting in complete repair, survival with mutations, or cell death. The study of UV-induced DNA damage and its repair provides a useful model to examine the role of NER in mutagenesis and carcinogenesis, due to our relatively complete understanding of the lesions and the repair mechanisms involved. The clinical, biochemical, and molecular evidence presented regarding diseases caused by deficiencies in NER, such as XP, clearly indicate that UV irradiation directly contributes to skin cancers in humans. However, a close analysis of the specific NER pathways which result in heterogenous repair of DNA, and the specific clinical phenotype resulting from mutations in the various genes involved, reveal a more complex relationship between repair and carcinogenesis. In fact, it appears from the study of patients with CS that defects in mechanisms for the preferential repair of transcribed DNA sequences do not result in a propensity to develop carcinogen-induced malignancies, but rather display clinical features that may be due to resulting abnormalities in transcription. In contrast, molecular defects in the repair of UV-induced DNA damage in the overall genome correlate closely with the dramatically increased incidence of skin cancers in XP patients. Thus, the recent discoveries at a molecular level of the various gene products and pathways involved in NER, and their interactions with other cellular processes, such as transcription, has greatly expanded our understanding of the pathogenesis of the clinically heterogeneous set of human diseases characterized by deficiencies in NER. Furthermore, analysis of the molecular defects underlying the different phenotypic presentations of these diseases allows insight into the repair processes important for affecting mutagenesis and the development of cancers.

Acknowledgements. We thank A.K. Ganesan and C.A. Smith for helpful discussions and constructive comments on the manuscript. JMF is a Howard Hughes Medical Institute Physician Postdoctoral Fellow and PCH is supported by Outstanding Investigator Award CA 44349 from the National Cancer Institute.

References

Bohr VA, Smith CA, Okumoto DS, Hanawalt PC (1985) DNA repair in an active gene: removal of pyrimidine dimers from the DHFR gene of CHO cells is much more efficient than in the genome overall. Cell 40: 359–369

Bootsma D, Hoeijmakers JHJ (1993) Engagement with transcription. Nature 363: 114–115

Boyce R, Howard-Flanders P (1964) Release of UV light-induced thymidine dimers from DNA in E. coli. Proc Natl Acad Sci USA 51: 293–300

Brash DE (1988) UV mutagenic photoproducts in E. coli and human cells: a molecular genetics perspective on human skin cancer. Photochem Photobiol 48: 59–66

Brash DE, Rudolph JA, Simon JA, Lin A, McKenna GJ, Baden HP, Halperin AJ, Ponten J (1991) A role for sunlight in skin cancer: UV-induced p53 mutations in squamous cell carcinoma. Proc Natl Acad Sci USA 88: 10124–10128

Bredberg A, Kraemer KH, Seidman MM (1986) Restricted ultraviolet mutational spectrum in a shuttle vector propagated in xeroderma pigmentosum cells. Proc Natl Acad Sci USA 83: 8273–8277

Bronner CE, Baker SM, Morrison PT, Watten B, Smith LG, Lescoe MK, Kane M, Earabino C, Lipford J, Kindblom A, Tannergard P, Bollag RJ, Godwin AR, Ward DC, Nordenskjold M, Fishel R, Kolodner R, Liskay RM (1994) Mutations in the DNA mismatch repair gene homologue hMLH1 is associated with hereditary non-polyposis colon cancer. Nature 368: 258–261

Cairns J (1981) The origin of human cancers. Nature 289: 353-357

Chen RH, Maher VM, Brouwer J, Van de Putte P, McCormick JJ (1992) Preferential repair and strand-specific repair of benzo[a]pyrene diol epoxide adducts in the HPRT gene of diploid human fibroblasts. Proc Natl Acad Sci USA 89: 5413–5417

Christians FC, Hanawalt PC (1992) Inhibition of transcription and strand-specific DNA repair by α-amanatin in Chinese hamster ovary cells. Mutat Res 274: 93–101

Christians FC, Hanawalt PC (1993) Lack of transcription-coupled repair in mammalian ribosomal RNA genes. Biochemistry 32: 10512–10518

Christians FC, Hanawalt PC (1994) Repair in ribosomal RNA genes is deficient in xeroderma pigmentosum group C and in Cockayne's syndrome cells. Mutat Res 323: 179–187

Cleaver JE (1968) Defective repair replication of DNA in xeroderma pigmentosum. Nature 218: 652–656

Cleaver JE, Kraemer KH (1989) Xeroderma pigmentosum. In: Scriver CR, Beudet AL, Sly WS, Valle D (eds.) Metabolic basis of inherited disease. McGraw-Hill, New York, pp 2949–2971

Donahue BA, Yin S, Taylor JS, Reines D, Hanawalt PC (1994) Transcript cleavage by RNA polymerase II arrested by a cyclobutane pyrimidine dimer in the DNA template. Proc Natl Acad Sci USA 91: 8502–8506

Drapkin R, Reardon JT, Ansari A, Huang JC, Zawel L, Ahn K, Sancar A, Reinberg D (1994) Dual role of TFIIH in DNA excision repair and in transcription by RNA polymerase II. Nature 368: 769–772

Dumaz N, Drougard C, Sarasin A, Daya-Grosjean L (1993) Specific UV-induced mutation spectrum in the p53 gene of skin tumors from DNA-repair-deficient xeroderma pigmentosum patients. Proc Natl Acad Sci USA 90: 10529–10533

Dumaz N, Stary A, Soussi T, Daya-Grosjean L, Sarasin A (1994) Can we predict solar ultraviolet radiation as the causal event in human tumours by analysing the mutation spectra of the p53 gene? Mutat Res 307: 375–386

Evans MK, Robbins JH, Ganges MB, Tarone RE, Nairn RS, Bohr VA (1993a) Gene-specific DNA repair in xeroderma pigmentosum complementation groups A, C, D, and F. J Biol Chem 268: 4839–4847

Evans MK, Taffe BG, Harris CC, Bohr VA (1993b) DNA strand bias in the repair of the p53 gene in normal and xeroderma pigmentosum group C fibroblasts. Cancer Res 53: 5377–5381
Fishel R, Lescoe MK, Rao MRS, Copeland NG, Jenkins NA, Garber J, Kane M, Kolodner R (1993) The human mutator gene homolog MSH2 and its association with hereditary nonpolyposis colon cancer. Cell 75: 1027–1038
Ford JM, Lommel L, Hanawalt PC (1994) Preferential repair of ultraviolet light-induced DNA damage in the transcribed strand of the human p53 gene. Mol Carcinogen 10: 105–109
Friedberg EC (1985) DNA repair. Freeman, New York
Friedberg EC, Bardwell AJ, Bardwell L, Wang Z, Dianov G (1994) Transcription and nucleotide excision repair – reflections, considerations and recent biochemical insights. Mutat Res 307: 5–14
Hanawalt PC (1993) Transcription-dependent and transcription-coupled DNA repair responses. In: Bohr AV, Wassermann K, Kraemer KH (eds) Proceedings of the Alfred Benzon symposium 5: DNA repair mechanisms. Munksgaard, Copenhagen, pp 231–242
Hanawalt P, Mellon I (1993) Stranded in an active gene. Curr Biol 3: 67–69
Hanawalt PC, Sarasin A (1986) Cancer-prone hereditary diseases with DNA processing abnormalities. Trends Genet 2: 188–192
Hanawalt PC, Donahue BA, Sweder KS (1994) DNA repair and transcription: collision or collusion? Curr Biol 4: 518–521
Henning KA, Li L, Iyer N, McDaniel LD, Reagan MS, Legerski R, Schultz RA, Stefanini M, Lehmann AR, Mayne LV, Friedberg EC (1995) The Cockayne syndrome group A gene encodes a WD repeat protein that interacts with CSB protein and a subunit of RNA polymerase II TFIIH. Cell 82: 555–564
Hoeijmakers JHJ (1993a) Nucleotide excision repair, I. From E. coli to yeast. Trends Genet 9: 173–177
Hoeijmakers JHJ (1993b) Nucleotide excision repair, II. From yeast to mammals. Trends Genet 9: 211–217
Hollstein M, Sidranksky D, Vogelstein B, Harris CC (1991) p53 mutations in human cancers. Science 253: 49–53
Kanjilal S, Pierceall WE, Cummings KK, Kripke ML, Ananthaswamy HN (1993) High frequency of p53 mutations in ultraviolet radiation-induced murine skin tumors: evidence for strand bias and tumor heterogeneity. Cancer Res 53: 2961–2964
Kantor GJ, Barsalou LS, Hanawalt PC (1990) Selective repair of specific chromatin domains in UV-irradiated cells from xeroderma pigmentosum complementation group C. Mutat Res 235: 171–180
Kraemer KH, Slor H (1985) Xeroderma pigmentosum. Clin Dermatol 3: 33–69
Kraemer KH, Lee MM, Scotto J (1984) DNA repair protects against cutaneous and internal neoplasia: evidence from xeroderma pigmentosum. Carcinogenesis 5: 511–514
Kress S, Sutter C, Strickland PT, Makhtar H, Schweizer J, Schwarz M (1992) Carcinogen-specific mutational pattern in the p53 gene in ultraviolet B radiation-induced squamous cell carcinomas of mouse skin. Cancer Res 52: 6400–6403
Leach FS, Nicolaides NC, Popadopoulos N, Liu B, Hen J, Parsons R, Peltomaki P, Sistonen P, Aaltonen LA, Nystrom-Lahti M, Guan WY, Zhang J, Meltzer PS, Yu JW, Kao FT, Chen D, Cerosaletti KM, Fournier REK, Todd S, Lewis T, Leach RJ, Naylor SL, Weissenbach J, Mecklin JP, Jarvinen H, Petersen GM, Hamilton SR, Green J, Jass J, Watson P, Lynch HT, Trent JM, de la Chappelle A, Kinzler KW, Vogelstein B (1993) Mutations of a mutS homolog in hereditary nonpolyposis colorectal cancer. Cell 75: 1215–1225
Leadon SA, Lawrence DA (1991) Preferential repair of DNA damage on the transcribed strand of the human metallothionein genes requires RNA polymerase II. Mutat Res 255: 67-78
Leadon SA, Lawrence DA (1992) Strand-selective repair of DNA damage in the yeast GAL7 gene requires RNA polymerase II. J Biol Chem 267: 23175–23182
Ljungman M, Zhang F (1996) Blocked RNA polymerase as a trigger for UV light-induced apoptosis. Oncogene (in press).
Loeb LA (1991) Mutator phenotype may be required for multistage carcinogenesis. Cancer Res 51: 3075–3079
Lommel L, Hanawalt PC (1993) Increased UV resistance of a xeroderma pigmentosum revertant cell line is correlated with selective repair of the transcribed strand of an expressed gene. Mol Cell Biol 13: 970–976
Ma L, Westbroek A, Jockemsen AG, Weeda G, Bosch A, Bootsma D, Hoeijmakers JHJ, van der Eb AJ (1994) Mutational analysis of ERCC3, which is involved in DNA repair and transcription initiation: identification of domains essential for the DNA repair function. Mol Cell Biol 14: 4126–4134

Malkin D, Li FP, Strong LC, Fraumeni JF, Nelson CE, Kim DH, Kassel J, Gryka MA, Bischoff JZ, Tainsky MA, Friend SH (1990) Germ line p53 mutations in a familial syndrome of breast cancer, sarcomas and other neoplasms. Science 250: 1233–1238
Mansbridge JN, Hanawalt PC (1983) Domain-limited repair of DNA in ultraviolet irradiated fibroblasts from xeroderma pigmentosum complementation group C. In: Friedberg EC, Bridges BA (eds) Cellular responses to DNA damage. Liss, New York, pp 195–207
Mayne LV, Lehmann AR (1982) Failure of RNA synthesis to recover after UV-irradiation: an early defect in cells from individuals with Cockayne's syndrome and xeroderma pigmentosum. Cancer Res 42: 1473–1478
Mayne LV, Lehmann AR, Waters R (1982) Excision repair in Cockayne syndrome. Mutat Res 106: 179–189
McGregor WG, Chen RH, Lukash L, Maher VM, McCormick JJ (1991) Cell cycle-dependent strand bias for UV-induced mutations in the transcribed strand of excision repair-proficient human fibroblasts but not in repair-deficient cells. Mol Cell Biol 11: 1927–1934
Mellon I, Hanawalt PC (1989) Induction of the Escherichia coli lactose operon selectively increases repair of its transcribed DNA strand. Nature 342: 95–98
Mellon I, Bohr VA, Smith CA, Hanawalt PC (1986) Preferential DNA repair of an active gene in human cells. Proc Natl Acad Sci USA 83: 8878–8882
Mellon I, Spivak G, Hanawalt PC (1987) Selective removal of transcription-blocking DNA damage from the transcribed strand of the mammalian DHFR gene. Cell 51: 241–249
Miller CW, Simon K, Aslo A, Kok K, Yokota J, Buys CHCM, Terada M, Koeffler HP (1992) p53 mutations in human lung tumors. Cancer Res 52: 1695–1698
Moles JP, Moyret C, Guillot B, Jeanteur P, Guilhou JJ, Theillet C, Basset-Seguin N (1993) p53 gene mutations in human epithelial skin cancers. Oncogene 8: 583–588
Mullenders LHF, Vrieling H, Venema J, van Zeeland AA (1991) Hierarchies of DNA repair in mammalian cells: biological consequences. Mutat Res 250: 223–228
Muriel WJ, Lamb JR, Lehmann AR (1991) UV mutation spectra in cell lines from patients with Cockayne's syndrome and ataxia telangiectasia, using the shuttle vector pZ189. Mutat Res 254: 119–123
Nakazawa H, English D, Randell PL, Nakazawa K, Martel N, Armstrong BK, Yamasaki H (1994) UV and skin cancer: specific p53 gene mutation in normal skin as a biologically relevant exposure measurement. Proc Natl Acad Sci USA 91: 360–364
Nance MA, Berry SA (1992) Cockayne syndrome: review of 140 cases. Am J Med Gen 42: 68–84
Papadopoulos N, Nicolaides NC, Wie YF, Ruben SM, Carter KC, Rosen CA, Haseltine WA, Fleischmann RD, Fraser CM, Adams MD, Venter JC, Hamilton SR, Petersen GM, Watson P, Lynch HT, Peltomaki P, Mecklin J, de la Chapelle A, Kinzler KW, Vogelstein B (1994) Mutation of a mutL homolog in hereditary colon cancer. Science 263: 1625–1629
Parris CN, Kraemer KH (1993) Ultraviolet-induced mutations in Cockayne syndrome cells are primarily caused by cyclobutane dimer photoproducts while repair of other photoproducts is normal. Proc Natl Acad Sci USA 90: 7260–7264
Pettijohn D, Hanawalt PC (1964) Evidence for repair-replication of UV damage in bacteria. J Mol Biol 9: 395–402
Pierceall WE, Mukhopadhyay T, Goldberg LH, Ananthaswamy HH (1991) Mutations in the p53 tumor suppressor gene in human cutaneous squamous cell carcinomas. Mol Carcinog 4: 445–449
Puisieux A, Lim S, Groopman J, Oztuk M (1991) Selective targeting of p53 gene mutational hotspots in human cancers by etiologically defined carcinogens. Cancer Res 51: 6185–6189
Rady P, Scinicariello F, Wagner RF, Tyring SK (1992) p53 mutations in basal cell carcinomas. Cancer Res 82: 3804–3806
Robbins JH (1988) Xeroderma pigmentosum: defective DNA repair causes skin cancer and neurodegeneration. JAMA 260: 384–388
Robbins JH, Kraemer KH, Lutzner MA, Festoff BW, Coon HG (1974) Xeroderma pigmentosum: an inherited disease with sun sensitivity, multiple cutaneous neoplasms and abnormal repair. Ann Intern Med 80: 221–228
Sancar A, Tang MS (1993) Nucleotide excision repair. Photochem Photobiol 57: 905–921
Sato M, Nishigori C, Zghal M, Yagi T, Takebe H (1993) Ultraviolet-specific mutations in p53 skin tumors in xeroderma pigmentosum patients. Cancer Res 53: 2944–2949
Schaeffer L, Roy R, Humbert S, Moncollin V, Vermeulen W, Hoeijmakers JHJ, Chambon P, Egly JM (1993) DNA repair helicase: a component of BTF2 (TFIIH) basic transcription factor. Science 260: 58–63

Schaeffer L, Moncollin V, Roy R, Staub A, Mezzina M, Sarasin A, Weeda G, Hoeijmakers JH, Egly JM (1994) The ERCC2/DNA repair protein is associated with the class II BTF2/TFIIH transcription factor. EMBO J 13: 2388–2392

Schmickel RD, Chu EHY, Trosko JE (1977) Cockayne syndrome: a cellular sensitivity to ultraviolet light. Pediatrics 60: 135–139

Seetharam S, Protic-Sabljic M, Seidman MM, Kraemer KH (1987) Abnormal ultraviolet mutagenic spectrum in plasmid DNA replicated in cultured fibroblasts from a patient with the skin cancer-prone disease, xeroderma pigmentosum. J Clin Invest 80: 1613–1617

Seetharam S, Kraemer KH, Waters HL, Seidman MM (1991) Ultraviolet mutational spectrum in a shuttle vector propagated in xeroderma pigmentosum lymphoblastoid cells and fibroblasts. Mutat Res 254: 97–105

Selby CP, Sancar A (1991) Gene- and strand-specific repair in vitro: partial purification of a transcription-repair coupling factor. Proc Natl Acad Sci USA 88: 8232–8236

Selby CP, Sancar A (1993) Molecular mechanism of transcription-repair coupling. Science 260: 53–58

Selby CP, Witkin EM, Sancar A (1991) Escherichia coli mfd mutant deficient in "mutation frequency decline" lacks strand-specific repair: in vitro complementation with purified coupling factor. Proc Natl Acad Sci USA 88: 11574–11578

Setlow RB, Carrier W (1964) The disappearance of thymidine dimers from DNA: an error correcting mechanism. Proc Natl Acad Sci USA 51: 226–231

Smerdon MT, Thoma F (1990) Site-specific DNA repair at the nucleosome level in a yeast minichromosome. Cell 61: 675–684

Sweder KS (1994) Nucleotide excision repair in yeast. Curr Genet 27: 1–16

Sweder KS, Hanawalt PC (1992) Preferential repair of cyclobutane pyrimidine dimers in the transcribed DNA strand in yeast chromosomes and plasmids is dependent upon transcription. Proc Natl Acad Sci USA 89: 10696–10700

Sweder KS, Hanawalt PC (1993) Transcription-coupled DNA repair. Science 262: 439–440

Sweder KS, Hanawalt PC (1994) The COOH terminus of suppressor of stem loop (SSL2/RAD25) in yeast is essential for overall genomic exicision repair and transcription-coupled repair. J Biol Chem 269: 1852–1857

Takayama K, Salazar EP, Lehmann A, Stefanini M, Thompson LH, Weber CA (1995) Defects in the DNA repair and transcription gene ERCC2 in the cancer-prone disorder xeroderma pigmentosum group D. Cancer Res 55: 5656–5663

Tessman I (1976) A mechanism of UV reactivation. In: Bukhari A, Lungquist E (eds) Abstracts of the bacteriophage meeting. Cold Spring Harbor Laboratory Press, New York, pp 87

Timme TL, Moses RE (1988) Review: diseases with DNA damage-processing defects. Am J Med Sci 295: 40–48

Troelstra C, Adijk H, de Wit J, Westerweld A, Thompson LH, Bootsma D, Hoeijmakers JHJ (1990) Molecular cloning of the human excision repair gene ERCC6. Mol Cell Biol 10: 5806–5813

Troelstra C, van Gool A, de Wit J, Vermeulen W, Bootsma D, Hoeijmakers JHJ (1992) ERCC6, a member of a subfamily of putative helicases, is involved in Cockayne's syndrome and preferential repair of active genes. Cell 71: 939-953

van Hoffen A, Natarajan AT, Mayne LV, van Zeeland AA, Mullenders LHF, Venema J (1993) Deficient repair of the transcribed strand of active genes in Cockayne's syndrome cells. Nucleic Acids Res 21: 5890–5895

van Vuuren AJ, Vermeulen W, Ma L, Weeda G, Appeldoorn E, Jaspers NGJ, van der Eb AJ, Bootsma D, Hoeijmakers JHJ, Humbert S, Schaeffer L, Egly JM (1993) Correction of xeroderma pigmentosum repair defect by basal transcription factor BTF2. EMBO J 13: 1645–1653

Venema J, Mullenders LHF, Natarajan AT, van Zeeland AA, Mayne LV (1990a) The genetic defect in Cockayne syndrome is associated with a defect in repair of UV-induced DNA damage in transcriptionally active DNA. Proc Natl Acad Sci USA 87: 4707–4711

Venema J, van Hoffen A, Natarajan AT, van Zeeland AA, Mullenders LHF (1990b) The residual repair capacity of xeroderma pigmentosum complementation group C fibroblasts is highly specific for transcriptionally active DNA. Nucleic Acids Res 18: 443–448

Venema J, van Hoffen A, Karcagi V, Natarajan AT, van Zeeland AA, Mullenders LHF (1991) Xeroderma pigmentosum complementation group C cells remove pyrimidine dimers selectively from the transcribed strand of active genes. Mol Cell Biol 11: 4128–4134

Venema J, Barosava Z, Natarajan AT, van Zeeland AA, Mullenders LHF (1992) Transcription affects the rate but not the extent of repair of cyclobutane pyrimidine dimers in the human adenosine deaminase gene. J Biol Chem 267: 8852–8856

Vermeulen W, Scott RJ, Rodgers S, Muller HJ, Cole J, Arlett CF, Kleijer WJ, Bootsma D, Hoeijmakers JHJ, Weeda G (1994) Clinical heterogeneity within xeroderma pigmentosum associated with mutations in the DNA repair and transcription gene ERCC3. Am J Hum Genet 54: 191–200

Vos JMH, Wauthier EL (1991) Differential introduction of DNA damage and repair in mammalian genes transcribed by RNA polymerase-I and polymerase-II. Mol Cell Biol 11: 2245–2252

Vrieling H, van Rooijen ML, Groen NA, Zdzienicka MZ, Simons JWIM, Lohman PHM, van Zeeland AA (1989) DNA strand specificity for UV-induced mutations in mammalian cells. Mol Cell Biol 9: 1277–1283

Vrieling H, Venema J, van Rooyen ML, van Hoffen A, Menichini P, Zdzienicka MZ, Simons JWIM, Mullenders LHF, van Zeeland AA (1991) Strand specificity for UV-induced DNA repair and mutations in the Chinese hamster HPRT gene. Nucleic Acids Res 19: 2411–2415

Wang YC, Maher VM, Mitchell DL, McCormick JJ (1993) Evidence from mutation spectra that the UV hypermutability of xeroderma pigmentosum variant cells reflects abnormal, error-prone replication on a template containing photoproducts. Mol Cell Biol 13: 4276–4283

Weeda G, van Ham RCA, Vermeulen W, Bootsma D, van der Eb AJ, Hoeijmakers JHJ (1990) A presumed DNA helicase encoded by ERCC-3 is involved in the human repair disorders xeroderma pigmentosum and Cockayne's syndrome. Cell 62: 777–791

Weeda G, Hoeijmakers JHJ, Bootsma D (1993) Genes controlling nucleotide excision repair in eukaryotic cells. Bioessays 15: 249–258

Yagi T, Tatsumi-Miyajima J, Sata M, Kraemer KH, Takebe H (1991) Analysis of point mutations in an ultraviolet-irradiated shuttle vector plasmid propagated in cells from Japanese xeroderma pigmentosum patients in complementation groups A and F. Cancer Res 51: 3177-3182

Yagi T, Sato M, Tatsumi-Miyajima J, Takebe H (1992) UV-induced base substitution mutations in a shuttle vector plasmid propagated in group C xeroderma pigmentosum cells. Mutat Res 273: 213-220

Yamaizumi M, Sugano T (1994) UV induced nuclear accumulation of p53 is evoked through DNA damage of actively transcribed genes independent of the cell cycle. Oncogene 9: 2775-2784

Ziegler A, Leffell DJ, Kunala S, Sharma HW, Gailani M, Simon JA, Halperin AJ, Baden HP, Shapiro PE, Bale AE, Brash DE (1993) Mutation hotspots due to sunlight in the p53 gene of nonmelanoma skin cancers. Proc Natl Acad Sci USA 90: 4216-4220

Chromosome Instability Syndromes: Lessons for Carcinogenesis

M.S. Meyn

Departments of Genetics and Pediatrics, Yale University School of Medicine, 333 Cedar Street, New Haven, CT 06510, USA

1 Introduction

The ability to maintain genomic integrity in the face of DNA damage is critical for survival. Biological organisms are not merely passive targets of DNA-damaging agents but actively respond to DNA damage in a variety of ways, e.g., the SOS system in *Escherichia coli* (Battista et al. 1990). The means by which cells achieve this goal are complex and involve DNA repair, genetic recombination, alterations in the cell cycle, and programmed cell death. This article discusses the chromosome instability syndromes, human diseases in which these homeostatic processes have broken down, resulting in cancer, immunodeficiency, growth failure, neurologic abnormalities, and mutagen sensitivity. Genetic instability in these syndromes may result from changes in DNA topology, loss of cell cycle checkpoint control, or dysregulation of lymphokine-mediated signal-transduction pathways, illustrating the diverse nature of the cellular processes that can affect the integrity of the genome.

The chromosome instability syndromes are a group of autosomal recessive conditions that include ataxia telangiectasia (A-T), Fanconi anemia, Bloom syndrome, Werner syndrome (MONNAT 1992), Nijmegen breakage syndrome (WEEMAES et al. 1994), and ataxia pancytopenia (LI et al. 1981). Focusing on the first three conditions, this article summarizes their homozygote and heterozygote phenotypes with an emphasis on genetic instability and cancer. Previous models for these diseases are discussed and new hypotheses proposed to explain how the underlying molecular defects in these syndromes give rise to their phenotypic abnormalities. The involvement of these genes in inherited and sporadic cancers is reviewed in depth, followed by discussion of the implications that these diseases have for both understanding the basic biology of cancer and improving the treatment of cancer patients.

2 Chromosome Instability Syndromes

2.1 A-T Phenotype

The first published report of individuals with A-T was probably that of SULLABA and HENNER (1926), who described three adolescent siblings with progressive choreoathetosis and ocular telangiectasias, whom they thought represented a variant of Hunt's familial athetosis. Fifteen years later, LOUIS-BAR (1941) independently described a 9-year-old boy with cerebellar ataxia and telangiectasias, whom he proposed represented a new syndrome. The condition became known as Louis-Bar syndrome until the name ataxia telangiectasia was proposed by BODER and SEDGWICK (1957). The pleiotrophic phenotype of A-T homozygotes is described below.

2.1.1 Neurologic Abnormalities

Ataxia telangiectasia homozygotes have multiple neurologic problems (reviewed in SEDGWICK and BODER 1991; WOODS and TAYLOR 1992). Although they typically have no functional abnormalities in the 1st year of life, they gradually lose cerebellar function, resulting in progressive ataxia, dysarthric speech, ocular apraxia, drooling, and choreoathetoid movements. Most A-T patients are wheelchair bound by the end of the 1st decade of life due to cerebellar dysfunction, and older A-T patients may develop intellectual arrest. Their functional neurologic abnormalities are accompanied by a continual loss of neurons from the central nervous system. The cerebellum is particularly affected due to the cumulative effects of ongoing Purkinjie cell death.

2.1.2 Immunodeficiencies

Ataxia telangiectasia homozygotes express both humeral and cellular immune defects (PETERSON and GOOD 1968; STROBER et al. 1968). They have absent or hypoplastic thymuses, low numbers of circulating T cells and functional impairment of T cell mediated immunity (e.g., delayed hypersensitivity and proliferative responses). They also have a distinctive pattern of immunoglobulin (Ig) deficiencies, including low levels of IgA, IgE, IgG_2, and IgG_4. The combination of immunodeficiency and progressive loss of cerebellar function seen in A-T homozygotes makes aspiration pneumonia the leading cause of death in these children, whose median life expectancy was estimated in a 1986 survey to be ~17 years (MORRELL et al. 1986).

2.1.3 Malignancy

In a retrospective review of the first 101 known cases, BODER and SEDGWICK (1963) noted that lymphoreticular malignancy is the second most common cause of death in children with A-T. Since then there have been three separate large-scale surveys that illustrate the range of tumors in A-T homozygotes: (a) the Immunodeficiency Cancer Registry periodically has reported on those A-T patients with cancer that have been entered into this international registry (SPECTOR et al. 1982; HECHT and HECHT 1990); (b) MORRELL et al. (1986) described the incidence of cancer in 263 A-T homozygotes as part of a survey of A-T in the United States; and (c) SEDGWICK and BODER (1991) recently reviewed published autopsy reports that included 30 A-T homozygotes who had developed cancer. The following comments are based on these studies plus additional published case reports.

Cancer is a frequent complication of ataxia-telangiectasia. For example, in the 1986 MORRELL; et al. survey, ~20% of 263 A-T patients had developed cancer at the time of ascertainment. Malignancies were a contributing cause of death in almost half of published autopsies surveyed by SEDGWICK and BODER (1991). Recent estimates place the lifetime risk of developing cancer at 30%–40% (e.g., PETERSON et al. 1992). MORRELL et al. (1986) calculated that Caucasian A-T homozygotes had a 61-fold excess risk of developing cancer. Their risk for lymphoma was even higher: a 252-fold increase. In the same study, the relative risks for African-American A-T homozygotes were calculated to be 184-fold higher than controls for all cancers and 750-fold higher for lymphomas.

Published surveys and case reports have consistently shown an age effect on the tumor type seen in A-T homozygotes. Figure 1 plots malignancies in A-T patients by age at onset and type, using data from the Immunodeficiency Cancer Registry (SPECTOR et al. 1982). As Fig. 1 illustrates, there is a bias towards lymphomas and acute lymphocytic leukemias prior to age 10, while epithelial tumors tend to predominate beginning in adolescence. The slight fall off in epithelial tumors after age 20 seen in this and other surveys may be a reflection of the declining population of A-T homozygotes who survive past age 20 rather than a true drop in

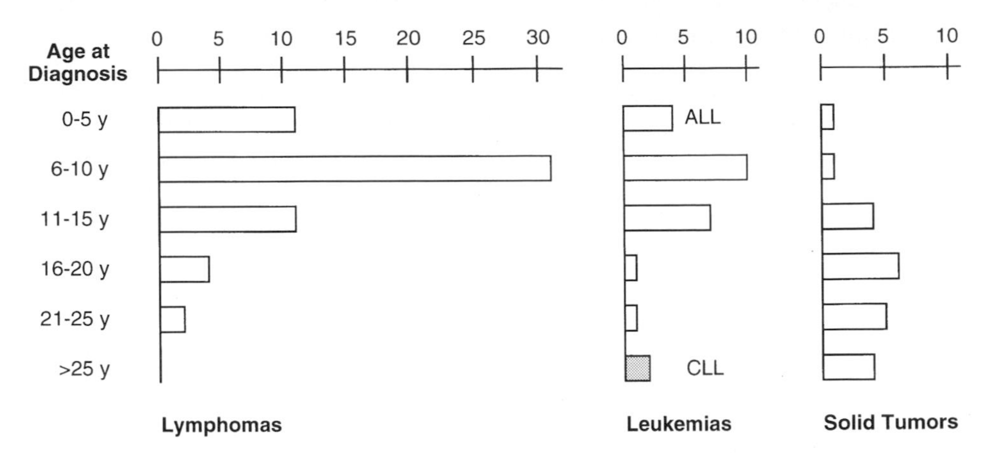

Fig. 1. Malignancies in A-T homozygotes as a function of age of onset (based on data from the Immunodeficiency Cancer Registry, SPECTOR et al. 1982)

the incidence of solid tumors in this age group. As noted by TAYLOR (1992), those few A-T homozygotes who live into their 20s and 30s are at risk to develop chronic T cell leukemias in addition to solid tumors (HECHT and HECHT 1990; SEDGWICK and BODER 1991; TAYLOR 1992).

The most common malignancies in A-T homozygotes are tumors of the immune system occurring in the first 15 years of life. For example, lymphomas and leukemias together account for 67% of the 52 tumors found in the 1986 MORRELL et al. survey and 74% of tumors seen in the 108 A-T patients listed in the Immunodeficiency Cancer Registry report of 1990 (HECHT and HECHT 1990). More than 40% of all tumors in A-T homozygotes are non-Hodgkin's lymphomas, another ~20% of tumors are acute lymphocytic leukemias (ALL) and ~5% are Hodgkin's lymphomas (MORRELL et al. 1986; HECHT and HECHT 1990; SEDGWICK and BODER 1991). Lymphomas in A-T homozygotes are commonly of B cell origin, while the majority of cases of ALL in A-T homozygotes express T cell markers in contrast to the low proportion of T cell-derived ALLs seen in other children (SPECTOR et al. 1982).

Solid tumors account for only about one quarter of tumors seen in A-T homozygotes. Nonetheless, they still are unusually frequent for such a young population. A wide range of cancers have been reported (SPECTOR et al. 1982; MORRELL et al. 1986; SEDGWICK and BODER 1991), the most frequent being gastric carcinoma, breast carcinoma, medulloblastoma, basal cell carcinoma, ovarian dysgerminoma, hepatoma, and uterine leiomyoma (in approximate order of frequency). Less common solid tumors include pyloric carcinoma, laryngeal carcinoma, gonadoblastoma, parotid carcinoma, thyroid carcinoma, renal cell carcinoma, pancreatic carcinoma, bladder carcinoma, epithelioma and leiomyosarcoma.

2.1.4 Other Clinical Abnormalities

Most A-T homozygotes develop telangiectasias of the sclera, face and ears, beginning at 3–6 years of age. Adolescent A-T homozygotes frequently have other cutaneous signs of premature aging, including greying of the hair, senile keratoses, skin atrophy, and areas of hyper- and hypopigmentation (SEDGWICK and BODER 1991). Gonadal abnormalities are another feature of the syndrome. Many female homozygotes have congenital absence or hypoplasia of the ovaries and menarche may be delayed or absent (SEDGWICK and BODER 1991; WOODS and TAYLOR 1992). Male homozygotes have been shown to have histological abnormalities of their testes, and incomplete spermatogenesis has been reported (STRICH 1963; AGUILAR et al. 1968), although hypogonadism is less frequent and milder than in female homozygotes. Some A-T homozygotes develop diabetes in adolescence, although this is not the norm (SCHALCH et al. 1970). Although older A-T patients develop growth retardation, congenital malformations are not a feature of A-T.

2.1.5 Genetic Instability in A-T Homozygotes

Genetic instability is an intrinsic feature of the A-T phenotype. The first direct evidence came from cytogenetic studies by HECHT et al. (1966), who found an increased frequency of chromatid and isochromatid breaks in PHA-stimulated lymphocytes from children with A-T. Since then, a variety of spontaneous in vivo chromosomal aberrations have been documented in multiple tissues from A-T homozygotes (reviewed in COHEN and LEVY 1989; KOJIS et al. 1991). The range of spontaneous aberrations includes chromatid and chromosome breaks, chromosome gaps, acentric fragments, and dicentric chromosomes, as well as increased frequencies of structural rearrangements and aneuploidy. Multiple cell lineages have been shown to express chromosomal instability, including T lymphocytes, keratinocytes, fibroblasts, hepatocytes, and neurons (e.g, HECHT et al. 1966; AGUILAR et al. 1968; KOJIS et al. 1991). However, most studies of bone marrow stem cells in A-T homozygotes have not documented increased chromosome aberrations (e.g., COHEN et al. 1975; KOHN et al. 1982b), and in vitro Epstein-Barr virus (EBV)-transformed lymphoblastoid lines do not express spontaneous karyotypic abnormalities (KOJIS et al. 1991). In addition, spontaneous sister chromatid exchange frequencies appear to be normal (GALLOWAY and EVANS 1975; BARTRAM et al. 1976; KOHN et al. 1982b). Interestingly, breakpoints have never been reported near the 11q23 location of the *ATM* gene itself. Telomere fusions, resulting in dicentric, multicentric, and ring chromosomes, are distinctive features of A-T homozygotes that are not seen in other chromosome instability syndromes. They occur in multiple cell types (PFEIFFER 1970; HAYASHI and SCHMID 1975; TAYLOR et al. 1981) and may be the result of shortened or abnormal telomeres (PANDITA et al. 1995).

In addition to the generalized increase in chromosome aberrations seen in other tissues from A-T homozygotes, their peripheral T lymphocytes have a specific predilection for rearrangements involving four sites: 7p14, 7q35, 14q11.2, and

14q32. Although translocations and inversions of chromosomes 7 and 14 also are common in normal individuals (AURIAS et al. 1985b; AURIAS and DUTRILLAUX 1986), they occur 30- to 50-fold more often in A-T homozygotes (AURIAS et al. 1980). These breakpoint regions contain the α, β, and γ T cell antigen receptor genes as well as the Ig heavy chain genes, immune genes that normally undergo obligatory rearrangement during the maturation of the immune system. When these breakpoints have been sequenced in A-T patients' T cells, they have interrupted their respective TCR and Ig heavy chain genes (reviewed in KOJIS et al. 1991). These T cell-specific rearrangements frequently exhibit clonality, and, with time, T cell clones carrying these chromosome 14 rearrangements can become the majority of peripheral T cells in A-T homozygotes (HECHT et al. 1973; OXFORD et al. 1975; SAADI et al. 1980). Although these rearrangements do not appear to be sufficient for malignant transformation, T cell leukemias in A-T homozygotes frequently arise from these clones (TAYLOR 1992). Inversions and 14 : 14 translocations involving the α-TCR and Ig heavy chain genes are common in chronic T cell leukemia in normal individuals as well (RABBITTS 1991), suggesting that these rearrangements may facilitate the development of T cell leukemia.

Spontaneous genetic instability also has been documented by molecular and immunologic means. A study of individuals heterozygous for the *M* and *N* alleles of the glycophorin locus found 7- to 14-fold elevations in frequencies of conversion to hemizygosity and homozygosity in erythrocytes from A-T homozygotes (BIGBEE et al. 1989). A-T homozygotes have approximately 70-fold elevation in their frequency of circulating T lymphocytes expressing γ/β, δ/β, α/δ, or γ/δ heavy chain TCRs as a result of aberrant interlocus gene rearrangements (LIPKOWITZ et al. 1990, 1992; KOBAYASHI et al. 1991). Their T lymphocytes also have significant elevation in the spontaneous frequency of *HPRT* mutations (COLE and ARLETT 1994). Finally, A-T fibroblast lines grown in culture have 30- to 200-fold increases in spontaneous intrachromosomal recombination rates (MEYN 1993).

2.1.6 Abnormal Responses to DNA Damage

In 1967, GOTOFF et al. described a 10-year-old A-T patient who developed lymphosarcoma and subsequently died from a severe reaction to radiation therapy directed at the tumor. This case report was the first of several that documented the in vivo sensitivity of A-T patients to the killing effects of ionizing radiation (see also MORGAN et al. 1968; FEIGIN et al. 1970; EYRE et al. 1988). In 1975 TAYLOR et al. confirmed this sensitivity in A-T fibroblasts grown in vitro. Since then, radiosensitivity has been documented for a variety of cell types, including primary fibroblasts, lymphocytes and keratinocytes as well as transformed lymphoblasts and fibroblasts (e.g., ARLETT et al. 1988; COLE et al. 1988; STACEY et al. 1989). Cells from A-T homozygotes are typically three to five times more sensitive to ionizing radiation than control cells, based on D_0 and D_{37} values (ARLETT et al. 1988; COLE et al. 1988). A-T cells are easily killed by other agents that induce double-strand breaks, particularly those that induce broken ends with 3′ phosphoglycolates (SHILOH et al. 1982a; BURGER et al. 1994; NELSON and KASTAN 1994). Most studies have found that A-T cells have normal resistance to UV irradiation (e.g., HEDDLE et al. 1983), suggesting that their sensitivity may be specific for double-strand breaks and gaps.

Another part of A-T radiosensitivity is an increase in the number of chromosomal aberrations induced by ionizing radiation or radiomimetic agents. Following treatment with agents that induce double-strand breaks, cells from A-T homozygotes express significant increases in the frequencies of induced chromatid breaks, gaps, and rearrangements (reviewed in COHEN and LEVY 1989). This has been documented in peripheral T cells, lymphoblastoid lines, primary fibroblasts, and fibroblast lines (COHEN and LEVY 1989). These increases in induced aberrations are not due to an increased susceptibility to the induction of DNA breaks by ionizing radiation (CORNFORTH and BEDFORD 1985; PANDITA and HITTELMAN 1992b).

2.1.7 Repair Abnormalities in A-T Homozygotes

Although cells from A-T homozygotes are exquisitely sensitive to the cytotoxic effects of ionizing radiation, their ability to repair DNA damage appears to be largely intact. Multiple biochemical studies have failed to detect gross abnormalities in the kinetics of single-strand and double-strand break repair in A-T cells (e.g., TAYLOR et al. 1975; LEHMAN and STEVENS 1977). Other reports have found no evidence that A-T cells are functionally defective in DNA repair (MURIEL et al. 1991; EADY et al. 1992). On the other hand, A-T homozygotes may have subtle defects in their ability to repair DNA breaks. Several studies found slight increases in the fraction of DNA breaks left unrepaired in irradiated A-T cells (CORNFORTH and BEDFORD 1985; BLOCHER et al. 1991), as well as abnormalities in the rejoining of restriction enzyme breaks in plasmids transfected into A-T fibroblasts (COX et al. 1984; DEBENHAM et al. 1988) This may be the result of impaired accuracy in strand rejoining (COX et al. 1984; DEBENHAM et al. 1988) or inability to repair a small but critical fraction of double-strand breaks (TAYLOR 1978; CORNFORTH and BEDFORD 1985; TAYLOR et al. 1989; PANDITA and HITTELMAN 1992a).

The subtle repair defects observed in A-T homozygotes may be the result of their inability to activate DNA repair following DNA damage. Exposing normal human cells to low doses of radiation prior to infection of irradiated virus improves viral survival and increases the number of mutant viruses recovered (SUMMERS et al. 1985). These effects, termed enhanced survival and enhanced mutagenesis, have been demonstrated using both single-stranded and double-stranded DNA viruses (SUMMERS et al. 1985; JEEVES and RAINBOW 1986; HILGERS et al. 1989) and is presumed to be the result of activation of one or more DNA repair processes by DNA damage. Although not extensively studied, damage-activated DNA repair appears to be impaired in A-T homozygotes, since A-T fibroblasts have been shown to lack the enhanced survival and enhanced mutagenesis expressed by control human cells for irradiated H-1 parvovirus and adenovirus 2 (JEEVES and RAINBOW 1986; HILGERS et al. 1987; BENNETT and RAINBOW 1988).

2.1.8 A-T Homozygotes Lack Cell Cycle Checkpoints

There are at least three damage-sensitive cell cycle checkpoints in mammalian cells: one at the G_1/S border, another in S phase, and one at the G_2/M boundary. In normal cells, these checkpoints are triggered following the induction of strand breaks in cellular DNA by a variety of agents (reviewed in MURRAY 1992; HARTWELL and KASTAN 1994). These checkpoints also may restrain the cell cycle temporarily in response to the generation of strand breaks, shortened telomeres, and other DNA damage that occurs spontaneously during the course of normal DNA metabolism (e.g., site-specific gene rearrangements, genetic recombination, and repair of replication errors).

The tumor suppressor protein p53 plays a key role in activating the G_1/S checkpoint following certain types of DNA damage (reviewed in HARTWELL and KASTAN 1994). G_1/S cell cycle arrest occurs via p53-mediated transcriptional activation of the *p21(WAF-1/CIP1/SDI1)* gene, which codes for a protein that binds to Cdk-cyclin complexes and inhibits their kinase activities (KASTAN et al. 1992; DULIC et al. 1994; EL-DEIRY et al. 1994). The S phase and G_2/M damage-activated checkpoints may be p53 independent (KOMATSU et al. 1989; RUSSELL et al. 1995); however, relatively little is known about their genetics and biochemistry.

Ataxia telangiectasia homozygotes lack the p53-mediated G_1/S damage-sensitive checkpoint, and the kinetics of p53, p21, and GADD45 induction by ionizing radiation are abnormal in the cells of A-T homozygotes (KASTAN et al. 1992; KHANNA and LAVIN 1993; CANMAN et al. 1994). A-T cells do not have the S phase checkpoint, since they fail to arrest DNA synthesis when irradiated in S phase, resulting in the phenomenon of radioresistant DNA synthesis (PAINTER and YOUNG 1980). The G_2/M checkpoint also appears to be defective in A-T cells, in that both A-T fibroblasts and lymphocytes irradiated in G_2 fail to undergo the initial radiation-induced G_2/M delay seen in normal cells (ZAMBETI-BOSSELER and SCOTT 1980; RUDOLPH and LATT 1989; BEAMISH and LAVIN 1994).

2.1.9 Dysfunctional Apoptosis in A-T Homozygotes

Although researchers have begun only recently to focus on the possible role of programmed cell death in the phenotype of A-T homozygotes, several in vivo and in vitro findings suggest that inappropriate apoptosis is responsible for certain aspects of the disease. Histologic analyses of cerebella taken from A-T homozygotes at autopsy document a high frequency of abnormal Purkinje and granule cells that exhibit the highly condensed, pyknotic nuclei expected from programmed cell death in neurons (BODER and SEDGWICK 1958; AGAMANOLIS and GREENSTEIN 1979; AMROMIN et al. 1979). We recently demonstrated that fibroblasts and lymphoblasts undergo apoptotic death in culture following exposure to low radiation and streptonigrin doses that do not induce appreciable apoptosis in control cells (MEYN et al. 1994, submitted). This inappropriate apoptosis appears to be p53 mediated, in that it is suppressed in A-T fibroblasts whose p53 protein has been functionally inactivated by transfection with either a dominant-negative *p53* gene or a human papilloma virus E6 gene. In these experiments, transfection-induced loss of p53 function did not affect survival of control fibroblasts, but transfected A-T cells acquired near-normal resistance to ionizing radiation, suggesting that p53-mediated apoptosis is the major cause of radiosensitivity in A-T cells in culture (MEYN et al. 1994, submitted).

2.2 *ATM* Gene

In the summer of 1995, an intensive search for the genes responsible for A-T culminated with the positional cloning of the *ATM* (Ataxia Telangiectasia Mutated) gene by SAVITSKY et al. (1995). The *ATM* gene is conserved in vertebrates and codes for a 12-kb transcript that is abundantly expressed in multiple tissues in vivo. The carboxy terminus of the putative ATM protein is homologous to that of at least four checkpoint proteins from other organisms: *Drosophila melanogaster* MEI-41, *Schizosaccharomyces pombe* Rad3, *Saccharomyces cerevesiae* MEC1p, and *S. cerevesiae* TEL1p (GREENWELL et al. 1995; HARI et al. 1995; SAVITSKY et al. 1995). The region of strongest homology between these five proteins contains a phosphatidylinositol 3-kinase (PI3-kinase) domain, suggesting that proteins are involved in signal transduction.

The *ATM* gene also shares phenotypic similarities with these genes. Like A-T homozygotes, *mei-41, rad3*, and *mec1* mutants are X-ray sensitive and lack damage-induced cell cycle checkpoints (JIMENEZ et al. 1992; WEINERT et al. 1994; HARI et al. 1995). In addition, A-T, *mei-41, rad3*, and *tel1* homozygotes express increased chromosomal instability and have high spontaneous rates of mitotic recombination (PHIPPS et al. 1985; SEDGWICK and BODER 1991; MEYN 1993; GREENWELL et al. 1995; HARI et al. 1995). Taken together, the physical and phenotypic similarities between these checkpoint genes and *ATM* suggest that the normal function of ATM protein is to activate multiple cellular functions in response to spontaneous and induced DNA damage.

2.3 Damage Surveillance Network Model as a Working Hypothesis for *ATM* Function

In the past, abnormalities of DNA repair, genetic recombination, chromatin structure, and cell cycle checkpoint control all have been proposed as the underlying defect in A-T (for a review see MEYN 1995). However, with the cloning of *ATM* gene and the recognition of the putative ATM protein's similarity to other cell cycle checkpoint proteins, it now is assumed that A-T is caused by the failure of mutant *ATM* genes to play their proper role in activating cellular responses to DNA damage (ENOCH and NORBURY 1995; KASTAN 1995; MEYN 1995; SAVITSKY et al. 1995).

Several years ago, we proposed a model for A-T in which the A-T defect results in inability to activate a group of diverse cellular functions in response to DNA damage (MEYN et al. 1994; MEYN 1995). Figure 2a depicts such a damage response network in which the ATM gene product plays a critical role. In this model, the detection of certain types of spontaneous or induced DNA damage triggers a signal transduction network, resulting in the activation of a group of cellular functions that promote genetic stability by temporarily arresting the cell cycle and enhancing DNA repair. At the same time, the ATM-dependent network promotes cellular survival by inhibiting execution of damage-induced programmed cell death. In addition to the five responses illustrated in Fig. 2a, there also may be other, as yet undefined, ATM-dependent functions.

The primary abnormality in A-T homozygotes presumably creates a defect in this network that prevents the activation of these cellular functions in response to strand breaks, shortened telomeres, and other DNA lesions (Fig. 2b). This inability to respond to spontaneous and induced DNA damage results in increased genomic instability as well as in an unusually low threshold for triggering of p53-mediated apoptosis by otherwise non-lethal DNA damage. These abnormalities lead, in turn, to the multiple in vivo and in vitro abnormalities seen in A-T homozygotes (Fig. 2b).

The damage surveillance network (DSN) model offers a unifying explanation of how a single-gene defect can cause the pleiotropic phenotype seen in A-T homozygotes and explains several puzzling aspects of the disease. The model assumes that the enzymatic machinery for DNA repair and genetic recombination is essentially intact and emphasizes the contribution of defective cell cycle checkpoints to genetic instability and immune defects, two cardinal features of A-T. By ascribing the disruption of immunoglobulin switch recombination and TCR rearrangements to cell cycle checkpoint abnormalities and postulating that disruptions of immune gene rearrangements and of repair of spontaneous DNA

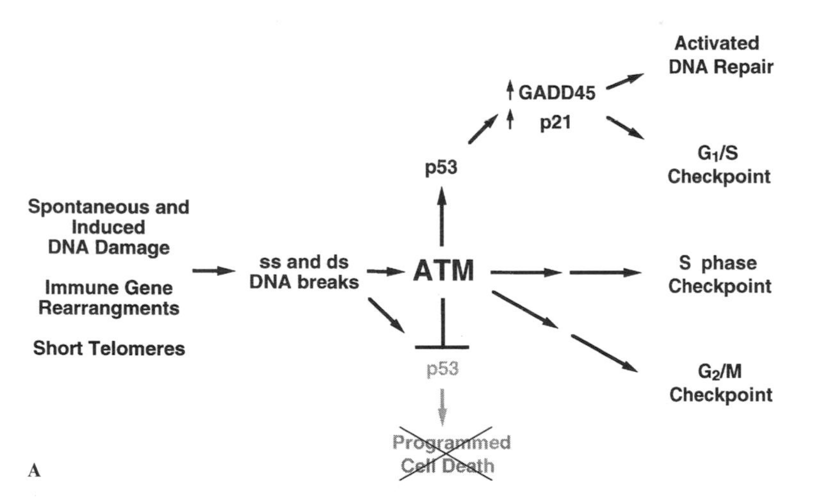

Fig. 2. A DNA damage surveillance network. As part of this signal transduction network, the ATM protein activates at least five cellular functions in response to the detection of spontaneous or induced DNA damage. **B** The DNA damage surveillance network is defective in A-T homozygotes. A-T homozygotes cannot activate ATM-dependent functions in response to DNA damage, resulting in the pleiotropic A-T phenotype

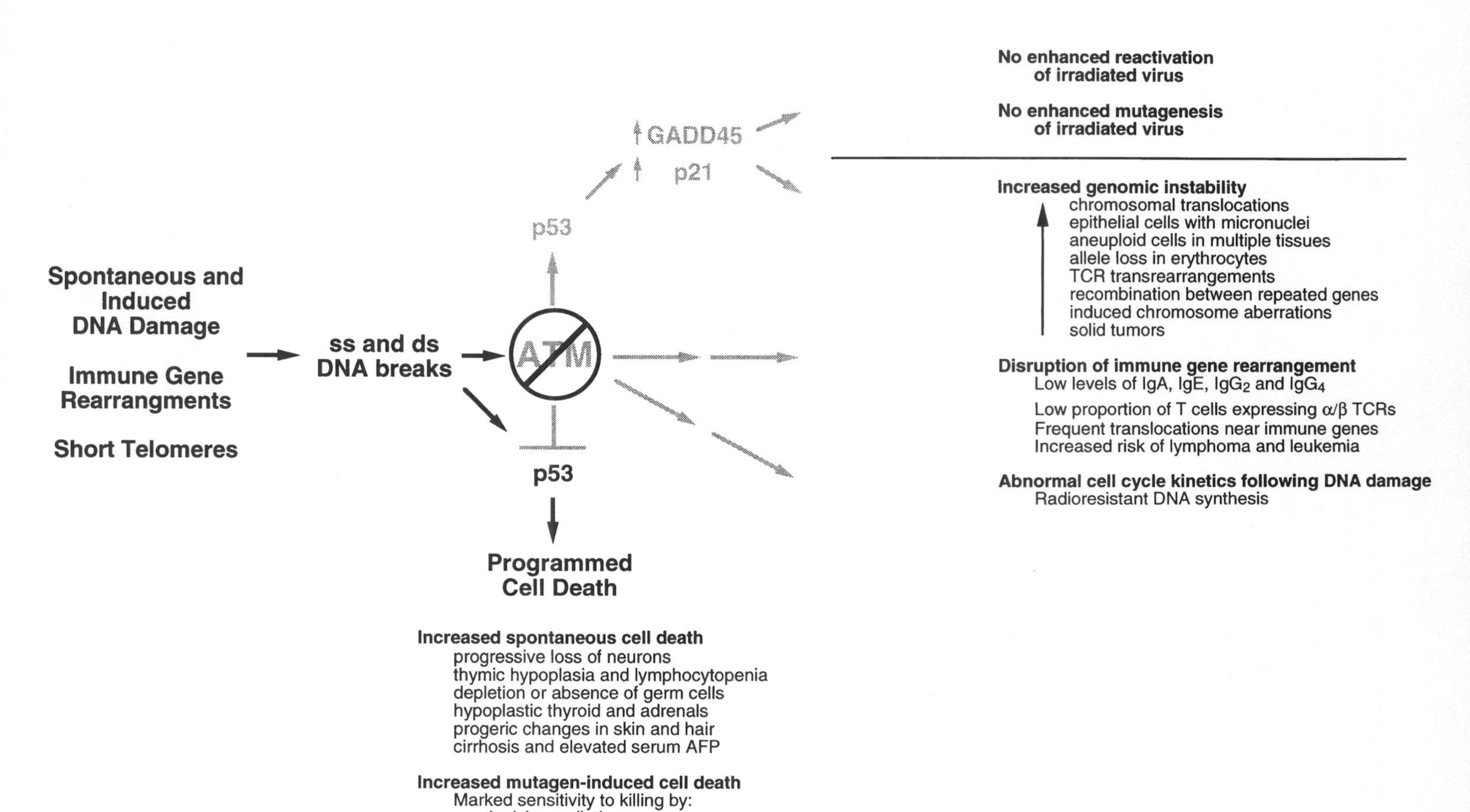

Spontaneous and Induced DNA Damage
Immune Gene Rearrangments
Short Telomeres
ss and ds DNA breaks
ATM
p53
GADD45
p21
p53
Programmed Cell Death
No enhanced reactivation of irradiated virus
No enhanced mutagenesis of irradiated virus
Increased genomic instability
chromosomal translocations
epithelial cells with micronuclei
aneuploid cells in multiple tissues
allele loss in erythrocytes
TCR transrearrangements
recombination between repeated genes
induced chromosome aberrations
solid tumors
Disruption of immune gene rearrangement
Low levels of IgA, IgE, IgG_2 and IgG_4
Low proportion of T cells expressing α/β TCRs
Frequent translocations near immune genes
Increased risk of lymphoma and leukemia
Abnormal cell cycle kinetics following DNA damage
Radioresistant DNA synthesis
Increased spontaneous cell death
progressive loss of neurons
thymic hypoplasia and lymphocytopenia
depletion or absence of germ cells
hypoplastic thyroid and adrenals
progeric changes in skin and hair
cirrhosis and elevated serum AFP
Increased mutagen-induced cell death
Marked sensitivity to killing by:
ionizing radiation
radiomimetic drugs

Fig. 2B

damage lead to generation of recombinogenic breaks and gaps in DNA, the DSN model explains why immunoglobulin switch recombination and TCR rearrangement appear to be defective in A-T homozygotes, while spontaneous rates of recombination between directly repeated non-immune genes in A-T fibroblasts are markedly higher than normal (MEYN 1993). The model also can account for the observation that chromosomal translocations in A-T lymphocytes cluster near immune genes, while translocations in A-T fibroblasts appear to involve random sites throughout the genome (KOJIS et al. 1989). By assuming that the ATM-dependent DSN normally monitors telomere integrity, the model explains telomeric abnormalities seen in A-T cells (PANDITA and HITTELMAN 1995).

The range of DNA lesions that might trigger the ATM network is uncertain, although strand breaks and gaps containing modified 3′ termini are likely to play a major role, given the sensitivity of A-T homozygotes to physical and chemical agents which induce strand breaks and small gaps containing 3′ phosphoglycolates (SHILOH et al. 1982a; BURGER et al. 1994; NELSON and KASTAN 1994). Short telomeres also may activate the ATM network, suggesting that the ends of abnormally short telomeres may be structurally similar to DNA breaks generated by these agents. As indicated in Fig. 2, the same lesions that activate A-T-dependent cellular functions also may trigger p53-mediated programmed cell death.

How far upstream of p53 the ATM protein functions in the signal transduction network is not certain. However, consideration of a related human protein, DNA-PK_{CS}, may be instructive in this regard. DNA-PK_{CS} is the catalytic subunit of DNA-PK, a DNA-dependent protein kinase (ANDERSON 1993). Ku70 and Ku80, the other subunits of DNA-PK, form a heterodimer that binds without sequence specificity to double-strand DNA breaks, gaps, and short hairpins (GOTTLIEB and JACKSON 1993). Once bound to DNA ends via the Ku polypeptides, DNA-PK activates its DNA-PK_{CS} subunit, which then can phosphorylate a variety of proteins in vitro, including p53 (ANDERSON 1993). This ability of the DNA-PK holoenzyme to phosphorylate proteins when bound to damaged DNA suggests that DNA-PK not only may have a direct role in promoting the repair of certain types of DNA damage, but also may serve as the front end to a signal transduction pathway that activates cellular responses to DNA damage. Further experimental support for a role in cellular damage responses for DNA-PK_{CS} is provided by recent evidence that a mutation in the *DNA-PK_{CS}* gene is responsible for the immune-deficient SCID mouse (BLUNT et al. 1995; KURCHGESSNER et al. 1995), and that mutant *DNA-PK_{CS}*, *ku70*, and *ku80* genes are associated with radiosensitivity, defects in double-strand break repair and abnormalities of VDJ recombination (TACCIOLI et al. 1994; TROELSTRA and JASPERS 1994; BUUL et al. 1995; LEES-MILLER et al. 1995). Although ATM and DNA-PK_{CS} proteins share strong sequence homology in their PI3-kinase domains (LEES-MILLER et al. 1995), their mutant phenotypes differ (e.g., Buul et al. 1995). In addition, A-T fibroblasts have normal intracellular amounts of the ku70, ku80, and DNA-PK_{CS} polypeptides and the DNA-PK enzymatic activity of A-T cell extracts is normal (LEES-MILLER et al. 1995). Taken together, these observations suggest that ATM and DNA-PK_{CS} act early and independently in separate damage response pathways. The similarities between ATM and DNA-PK_{CS} suggest that, like DNA-PK_{CS}, the ATM protein may be directly involved in the recognition of DNA damage, perhaps serving as the protein kinase subunit of a functional complex that also includes ku70- and ku80-like polypeptides.

It is generally assumed that a major function of ATM protein is signal transduction. It is not yet clear how this occurs. However, sequence analysis of the ATM protein indicates that it has a PI3-kinase domain (SAVITSKY et al. 1995). Similar PI3-kinase domains are found in the related proteins DNA-PK_{CS}, MEI-41, MEC1, RAD3, and TEL1 (GREENWELL et al. 1995; HARI et al. 1995; HARTLEY et al. 1995; SAVITSKY et al. 1995). The existence of a PI3-kinase domain in the ATM protein raises the possibility that a phosphoinositide might serve as a secondary messenger for the ATM signal transduction network. However, although the ATM protein has a PI3-kinase domain, its enzymatic activity is unknown, and it is far from certain that phosphoinositols are the biologically relevant targets for its putative phosphotransferase activity. The most closely related mammalian protein, DNA-PK_{CS}, has no detectable phosphinositol kinase activity in vitro, but can phosphorylate many proteins, including p53 (HARTLEY et al. 1995). Recently, it was demonstrated that, following irradiation, the *S. cerevisiae* ATM homologue Mec lp activates its downstream target RAD53p by phosphorylation, thereby triggering multiple cell cycle checkpoints and other cellular responses (SANCHEZ et al. 1996). The behavior of these two ATM homologues suggests that the true in vivo targets of the ATM protein's phosphotransferase activity might be one or more proteins. This conclusion is supported by the recent demonstration that ATM protein has in vitro protein kinase activity (KEEGAN et al. 1996). One target of a putative ATM kinase activity may be p53, since p53 has multiple phosphorylation sites and its DNA binding and transcriptional activities are activated by phosphorylation (reviewed in BOULIKAS 1995).

2.3.1 How Defects in Cellular Damage Responses Promote Cancer in A-T Homozygotes

The lack of multiple cell cycle checkpoints in the somatic cells of A-T homozygotes presumably increases the frequency of spontaneous and induced chromosome aberrations, mitotic recombination, and loss of heterozygosity. The result is a loss of control of genetic integrity and an unusually rapid accumulation of the multiple genetic changes in tumor suppressor genes and oncogenes necessary to give rise to tumors.

The highest cancer risk faced by A-T homozygotes is for leukemia and non-Hodgkin's lymphoma, tumors which frequently harbor chromosome rearrangements involving Ig supergene family genes. These rearrangements may be a result of loss of checkpoint control in response to a specific set of strand breaks, those that occur during the normal rearrangement and repair of immune gene DNA. The high proportion of lymphoid tumors seen in A-T homozygotes suggests that the initial production of strand breaks and other DNA damage is a rate-limiting step in oncogenesis, even in cells that are genetically unstable due to a lack of DNA damage-sensitive cell cycle checkpoints.

2.4 Fanconi Anemia Phenotype

In 1927 FANCONI published a report in which he described a new familial aplastic anemia (FANCONI 1927). Later, VAN LEEUVEN (1933) gave the name Fanconi anemia (FA) to this autosomal recessive syndrome of pancytopenia, multiple congenital malformations, malignancy, and genetic instability (for recent clinical reviews see ALTER 1994a; AUERBACH 1995). FA is the most common cause of inherited aplastic anemia, accounting for 10%–15% of all children with aplastic anemia (ALTER 1994b). FA homozygotes typically survive into their mid-20s, with death from bone marrow failure most common, followed by cancer (AUERBACH et al. 1989; ALTER 1994a).

2.4.1 Hematological Abnormalities

Fanconi anemia homozygotes have multiple hematological problems (reviewed in ALTER 1992, 1994a). The pancytopenia in FA is initially mild but relentlessly progressive, with an estimated 98% risk of bone marrow failure by age 40 (BUTTURINI et al. 1994). By the time FA patients are symptomatic, their bone marrow is hypocellular (CHU et al. 1979). Late in their pancytopenia, bone marrow cells from FA homozygotes can exhibit clonal chromosomal abnormalities (reviewed in BERGER and CONIAT 1989). The anemia responds to androgen therapy, but only bone marrow transplantation is curative (GLUCKMAN et al. 1995).

2.4.2 Malignancy

More than 40 years ago COWDELL et al. (1955) reported a 27-year-old man with FA who developed acute leukemia. Since then case reports and periodic surveys of the International Fanconi Anemia Registry (IFAR) (AUERBACH and ALLEN 1991; AUERBACH 1995) have documented that cancer is a common complication of FA. For example, in a recent survey of 836 published cases of FA, ALTER (1994a) found that 17% had developed cancer. The same survey found a shift towards early ages of onset and the occurrence of multiple tumors in 10% of FA homozygotes with cancer.

The most common malignancy in FA homozygotes is myeloid leukemia, which accounts for nearly half the malignancies seen in FA (ALTER 1994a). It has been estimated that the relative risk for developing myeloid leukemia in individuals with FA is 15 000-fold higher than normal (AUERBACH and ALLEN 1991), with 52% risk of developing myelodysplastic syndrome and/or acute myeloid leukemia by age 40 (BUTTURINI et al. 1994). Unlike A-T and Bloom syndrome (BS) lymphoma is uncommon in FA homozygotes, with only one case in 144 malignancies reported in the literature (ALTER 1994a).

Solid tumors account for half of the tumors seen in FA homozygotes and are unusually frequent for such a young population. A wide range of cancers is found (ALTER 1994a), the most frequent being hepatomas and hepatocellular carcinomas and oropharyngeal, gastrointestinal, and gynecological tumors (in approximate order of frequency).

Like A-T and BS, there is an age effect on the tumor type in FA homozygotes. Figure 3 plots malignancies (excluding liver tumors) in FA patients by age at onset and type, using data from the largest survey of the literature (ALTER 1994a). There is a bias towards leukemia prior to age 20, while solid tumors tend to predominate after age 25. The fall off in solid tumors after age 30 may be a reflection of the declining population of surviving FA homozygotes rather than a true drop in the incidence of solid tumors in older individuals.

Liver tumors account for 18% of malignancies in FA homozygotes. GERMAN (1983) suggested a possible link between cancer in FA and exposure to the anabolic steroids commonly used to treat anemia in these individuals. His original hypothesis that androgens play a major role in the development of *all* cancers in FA has not been supported by subsequent data. For example, among published cases, 45% of leukemias and 40% of solid tumors occurred in FA homozygotes who had never been exposed to androgens (ALTER 1994a). However, 94% of published cases of liver cancer in FA homozygotes occurred in individuals treated with androgens (ALTER 1994a), suggesting a specific link between liver cancer and androgens in these individuals.

2.4.3 Other Clinical Abnormalities

Fanconi anemia homozygotes "classically" have a variety of congenital malformations. However, 30% of patients in the 1989 IFAR survey had no physical

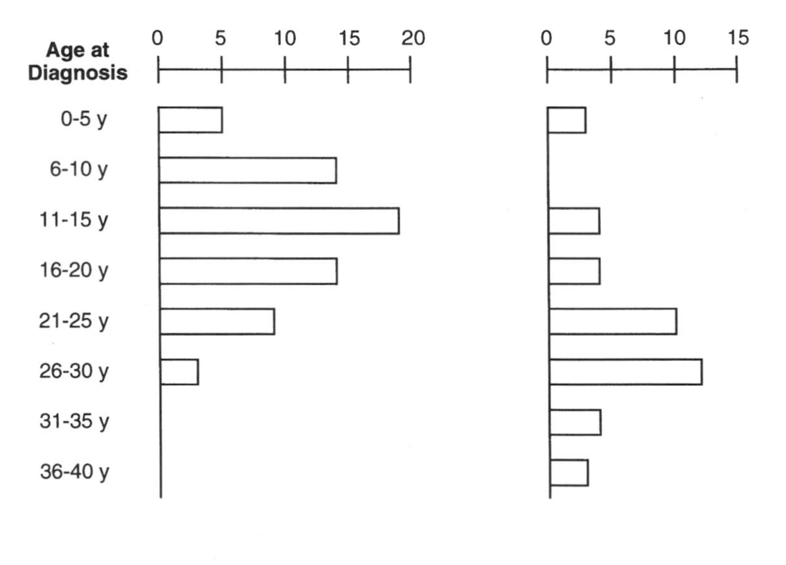

Fig. 3. Malignancies in FA homozygotes as a function of age of onset (based on the literature survey of ALTER 1994a)

malformations (AUERBACH et al. 1989), and in studies of the affected siblings of index cases, only ~25% had morphologic anomalies (RILEY et al. 1979; GLANZ and FRASER 1982). Almost half of FA homozygotes have some degree of growth retardation. Other common (i.e., >20% incidence) physical abnormalities include skin lesions (generalized hyperpigmentation, hypopigmented areas, and café-au-lait spots), facial dysmorphology (including microcephaly, frontal bossing, micrognathia), absent thumbs and/or other radial ray defects, structural abnormalities of the male genitals, and renal malformations (reviewed in ALTER 1994a; AUERBACH 1995). Less common morphologic abnormalities include vertebral anomalies, cardiac malformations, lower limb defects, deafness and malformations of the female genitourinary tract (ALTER 1994a; AUERBACH 1995). Like A-T and BS, gonadal abnormalities and reduced fertility are common problems in FA. Men with FA typically have azoospermia (BARGMAN et al. 1977), and female homozygotes undergo early menopause (ALTER 1994a). However, FA females have borne children, despite their anemia (ALTER 1994a). Unlike A-T, FA homozygotes do not appear to have intrinsic immune defects (ALTER 1992), although they occasionally have low serum levels of Ig (ALTER 1994a), and one report found decreased T cell function in vitro (JOHANSSON et al. 1982). Neurodevelopmental problems also have been reported in 25%–33% of FA homozygotes (PAVLAKIS et al. 1992; AUERBACH 1995).

2.4.4 Genetic Instability in FA

Genetic instability is a defining characteristic of FA. The first direct evidence for genetic instability in FA came from cytogenetic studies by SCHROEDER et al. (1964), who found an increased frequency of spontaneous chromosome aberrations in lymphocytes from children with FA. Since then, a variety of spontaneous in vivo chromosomal aberrations have been documented in circulating lymphocytes, fibroblasts, and bone marrow cells of FA homozygotes in vivo as well as in FA lymphocytes, fibroblasts, and lymphoblasts grown in culture (reviewed in SCHROEDER 1982; COHEN and LEVY 1989). The predominant and characteristic aberrations seen in FA homozygotes are single chromatid breaks and gaps and multiradials involving non-homologous chromosomes (SCHROEDER 1982; COHEN and LEVY 1989). Other abnormalities include endoreduplication (e.g., SCHROEDER et al. 1964) and the appearance of clonal populations of lymphocytes with marker chromosomes (HURET et al. 1988). Spontaneous sister chromatid exchange (SCE) frequencies are normal (e.g., LATT et al. 1975), but increased frequencies of spontaneous chromosome aberrations are nearly universal in FA homozygotes (SCHROEDER and KURTH 1971) and predate their development of anemia (PERKINS et al. 1969). Clones of bone marrow stem cells harboring chromosomal translocations do arise (e.g., HURET et al. 1988), but their role, if any, in bone marrow failure and the development of leukemia is not yet clear (see ALTER et al. 1993).

Several molecular studies have documented spontaneous genetic instability in FA homozygotes. SALA-TREPAT et al. (1993) found 30-fold elevations in the frequency of NO variants in erythrocytes from FA homozygotes and eight fold elevations in NN variant frequencies. BIGBEE et al. (1991) found similar elevations in glycophorin variant frequencies in nine FA homozygotes. VIJAYALAXMI et al. (1985) noted high spontaneous mutation frequencies at the *HPRT* locus in circulating T lymphocytes from FA homozygotes, although this result contrasts with that of SALA-TREPAT (1993), who found no significant differences between $HPRT^-$ mutant frequencies in lymphocytes from 23 FA homozygotes, when compared to 19 age-matched controls.

An unusually high proportion of spontaneous $HPRT^-$ mutants in the cells of FA homozygotes are deletions rather than point mutations (PAPADOPOULO et al. 1990a; LAQUERBE et al. 1995). LAQUERBE et al. (1995) sequenced the breakpoint junctions of 11 $HPRT^-$ deletion mutants in FA lymphoblasts and found that most of the deletion junctions involved a putative signal sequence similar to that used in VDJ recombination. Taken together, these results suggest that one consequence of the underlying defect in FA might be an inappropriate involvement of the cell's recombination machinery in the repair of spontaneous DNA lesions. As pointed out by SALA-TREPAT et al. (1993), the preferential elevation of glycophorin but not $HPRT^-$ variants in FA homozygotes is consistent with this possibility, as glycophorin variants frequently arise as a result of genetic recombination, while recombination does not usually play a major role in the repair of DNA damage at the hemizygous *HPRT* locus. This idea that the genetic recombination is involved in the genetic instability seen in FA homozygotes is further supported by our own observations that a group A FA fibroblast line grown in culture averaged 17-fold higher spontaneous intrachromosomal recombination rates than control human fibroblasts (MEYN et al. 1993).

2.4.5 Abnormal Responses to DNA Damage

The characteristic response of cells from FA homozygotes to many DNA-damaging agents is an unusually low threshold for the induction of chromosomal aberrations (reviewed in SCHROEDER 1982; COHEN and LEVY 1989). FA cells are particularly sensitive to the induction of clastogenic damage by bifunctional DNA cross-linking agents such as mitomycin C (MMC), diepoxybutane (DEB), UV-activated 8-MOP, *cis*-diaminedichloroplatinum (II), and HN_2 (SASAKI and TONOMURA 1973; LATT et al. 1975; AUERBACH and WOLMAN 1976; BERGER et al. 1980; POLL et al. 1985). For example, chromosomal breaks induced by 100 ng/ml DEB average 8.96 breaks/cell in peripheral lymphocytes from FA homozygotes compared to an average of 0.06 breaks/cell for controls (AUERBACH et al. 1989). DEB-induced chromosomal instability is such a consistent characteristic of FA that it is used as the diagnostic benchmark by the IFAR (AUERBACH 1993). Both spontaneous and induced aberrations in FA consist primarily of chromatid breaks (SCHROEDER and KURTH 1971; LATT et al. 1975; PARSHAD et al. 1983), suggesting the functional defect(s) in FA are manifest in G_2. In contrast to BS, induction of SCEs by many DNA-damaging agents is either normal or decreased in FA cells (LATT et al. 1975; PORFIRIO et al. 1983).

The sensitivity of FA cells grown in culture to the clastogenic effects of bifunctional DNA cross-linking agents is paralleled by a marked susceptibility to their cytotoxic effects, as demonstrated by the relative ease by which FA cells are killed by exposure to MMC, NH_2, DEB, and UV-activated 8-MOP (FUJIWARA and TATSUMI 1977; FORNACE et al. 1979; WEKSBERG et al. 1979; ISHIDA and BUCHWALD 1982; WUNDER and FLEISCHER-REISCHMANN 1983). Interestingly, FA cells have relatively normal resistance to clastogenic and cytotoxic effects of monofunctional DNA-alkylating agents such as 4NQO, MMS, MNNG, *trans*-diaminedichloroplatinum (II), and decarbamoyl-MMC (SASAKI and TONOMURA 1973; SASAKI 1978; POLL et al. 1985), suggesting a specific sensitivity to interstrand cross-links. Some, but not all studies, have reported that FA cells are sensitive to the clastogenic and cytotoxic effects of ionizing and UV radiation (HIGURASHI and CONEN 1971; SASAKI and TONOMURA 1973; FUJIWARA and TATSUMI 1977; SASAKI 1978; BIGELOW et al. 1979; ARLETT and HARCOURT 1980; DRITSCHILO et al. 1984; DUCKWORTH-RYSIECKI and TAYLOR 1985). Many FA homozygotes express an in vivo sensitivity to the toxic effects of X-irradiation and cyclophosphamide that are used to ablate bone marrow prior to transplantation (e.g., GLUCKMAN et al. 1983).

Despite their marked sensitivity for the induction of chromosome aberrations by DNA cross-linking agents, FA cells may be hypomutable at certain genomic loci. For example, exposure to UV-activated 8-MOP induced fewer mutations at the *HPRT* and Na^+/K^+-*ATPase* loci in FA(A) and FA(B) cells than in controls (PAPADOPOULO et al. 1990a,b)

2.4.6 Oxygen Sensitivity in FA

A large body of literature documents the in vitro sensitivity of FA cells to the toxic effects of oxygen and other reactive oxygen species (ROS) as well as the beneficial effects of antioxidant enzymes and low molecular weight antioxidants on the growth and chromosomal stability of FA cells (reviewed in ALTER 1994c; DEGAN et al. 1995). For example, growth in low oxygen concentrations decreases spontaneous chromosomal aberrations in FA cells (JOENJE et al. 1981) and corrects their slow growth and G_2 cell cycle abnormalities (SCHINDLER and HOEHN 1988). Addition of superoxide dismutase (SOD) or catalase to the media of FA lymphocytes can both reduce their spontaneous chromosome breakage (NORDENSON 1977) and increase their survival after exposure to MMC (NAGASAWA and LITTLE 1983), while the presence of low molecular weight antioxidants in the culture media partially protects FA lymphocytes against the clastogenic effects of DEB and H_2O_2 (DALLAPICCOLA et al. 1985).

The etiology of this sensitivity to reactive oxygen species is not clear. No consistent abnormalities have been identified in the major cellular pathways for detoxification of reactive oxygen species. In various reports, FA cells have been found to have either normal, decreased, or increased intracellular levels of SOD, catalase, and glutathione peroxidase (JOENJE et al. 1979; OKAHATA et al. 1980; MAVELLI et al. 1982; DALLAPICCOLA et al. 1984; YOSHIMITSU et al. 1984; SCARPA et al. 1985; GILLE et al. 1987). In the FA cells that have been examined, the specific activity and electrophoretic mobility of CuZn-SOD are normal (JOENJE et al. 1979), as are intracellular levels of glutathione (POOT et al. 1986).

Reactive oxygen species can be mutagenic as well as cause increased chromosome aberrations (Imlay and Linn 1988). Nordenson (1977) first postulated that ROS-induced DNA damage was responsible for spontaneous chromosomal instability in FA. Consistent with this view that FA cells are particularly vulnerable to ROS-induced DNA damage, Degan et al. (1995) found that DNA isolated from the circulating lymphocytes of FA homozygotes contains four- to fivefold more 8-hydroxy 2′-deoxyguanosine (a DNA-H_2O_2 reaction product) than DNA from control individuals. However, Seres and Fornace (1982) found that high levels of oxygen induced the same number of single-strand DNA breaks in FA and control cells, suggesting that, if ROS affect chromosomal instability in FA cells, their effect may not be mediated through direct ROS damage of DNA.

In order to account for the sensitivity of FA cells to both ROS and bifunctional cross-linking agents, it has been suggested that compounds such as MMC exert their genotoxic effects on FA cells via the production of oxygen-dependent free radical intermediates (Nagasawa and Little 1983; Pritsos and Sartorelli 1986). However, several observations suggest that this may not be the case. Although MMC and DEB may generate intracellular free radical intermediates, FA cells also are sensitive to the UV-activated 8-MOP, a DNA cross-linking agent whose action is thought to be relatively independent of ROS species (Averbeck and Averbeck 1985). In addition, oxygen toxicity and exposure to DNA cross-linking agents have different effects on the cell cycle in FA cells (Hoehn et al. 1989), and SV40 transformation of FA fibroblasts eliminates most of their oxygen sensitivity without affecting their sensitivity to the cytotoxic effects of MMC (Saito et al. 1993), demonstrating that these two aspects of the FA phenotype can be dissociated in certain circumstances.

2.4.7 Repair Abnormalities in FA Homozygotes

Sasaki (1975) first proposed that the underlying problem in FA is a defect in the repair of interstrand cross-links in DNA. Initial studies suggested problems with an early step in the removal of DNA interstrand cross-links, the endonucleolytic unhooking of the lesion from one strand (Sasaki 1975; Fujiwara and Tatsumi 1977; Fujiwara 1982). However, other reports found no defects in cross-link repair in FA cells (Fornace et al. 1979; Kaye et al. 1980; Poll et al. 1984). One potential explanation for these inconsistencies is that they reflect different complementation groups. In support of this possibility, Rousset et al. (1990) found that FA(A) cells had impaired recognition and/or removal of 8-MOP-induced cross-links while FA(B) cells were normal. In addition, Lambert et al. (1992) and Hang et al. (1993) have demonstrated that extracts from FA(A) cells, but not FA(B) cells, have somewhat less endonuclease activity than controls against psoralin-induced interstrand cross-links in vitro. Multiple enzyme defects have been proposed for FA (reviewed in Cohen and Levy 1989; Alter 1994c). However, except for the putative cross-link endonuclease deficiency found by Lambert et al. and Hang et al., those enzyme activities that have been examined have been normal (e.g., Teebor and Duker 1975; Willis and Lindahl 1987).

Studies that use viruses to probe the repair of DNA cross-links have shown that FA cells that are sensitive to psoralin and *cis*-platinum express normal host-cell reactivation of UV plus psoralin-treated herpes simplex virus (HALL and SCHERER 1981; FENDRICK and HALLICK 1984), UV plus psoralin-treated adenovirus (DAY et al. 1975), and *cis*-platinum-treated SV40 DNA (POLL et al. 1984). These results suggest that, despite FA cells' sensitivity to the cytotoxic and clastogenic effects of DNA cross-linking agents, their repair of interstrand cross-links in episomal DNA is normal, raising doubts as to whether there is a defect in the enzymatic machinery for DNA cross-link repair.

2.4.8 Cell Cycle in FA

Primary cultures of FA cells grow poorly in culture due to delayed progression through S and G_2 (SASAKI 1975; WEKSBERG et al. 1979; DUTRILLAUX et al. 1982). However, no defect in DNA replication has been documented. α, β and γ DNA polymerases appear to be normal in FA cells (BERTAZZONI et al. 1978), and rates of DNA chain elongation and fork displacements have been found to be the same as controls (HAND and GERMAN 1975; KAPP and PAINTER 1981). The etiology of this slow growth is not clear; however, it may be linked to ROS sensitivity, since low concentrations of ambient oxygen correct both the slow growth rate and unusually prolonged G_2 phase seen in FA cells grown in culture (SCHINDLER and HOEHN 1988).

The induction of p53 and activation of cell cycle checkpoints by DNA damage have not been examined extensively in FA cells. However, ROSSELLI et al. (1995) have demonstrated that, compared to normal cells, p53 is poorly induced by X-rays, MMC, and UV-B in group C and group D FA lymphoblasts. Surprisingly, they found that these same FA lymphoblasts had an intact X-ray induced G_1/S checkpoint, suggesting that induction of p53 and activation of the G_1/S checkpoint can be uncoupled under certain circumstances. Further evidence for abnormalities in the cell cycle checkpoint responses of FA cells is provided by DEAN and FOX (1983) who found that HN_2 exposure induced transient S phase delays in normal cells but not FA fibroblasts. FA cells treated with bifunctional cross-linking agents typically arrest in the first G_2 following exposure (DUTRILLAUX et al. 1982; KUBBIES et al. 1985; SEYSCHAB et al. 1994). This arrest can be overcome by treatment with caffeine (SEYSCHAB et al. 1994), a known inhibitor of the G_2 DNA damage-sensitive checkpoint, suggesting that the G_2/M checkpoint is intact in FA cells.

2.4.9 Apoptosis in FA Homozygotes

Concentrations of bifunctional cross-linking agents needed to induce even one to three chromosome breaks per cell are typically 100- to 1000-fold higher than those needed to effectively kill either normal or FA cells (ISHIDA and BUCHWALD 1982; WUNDER and FLEISCHER-REISCHMANN 1983), suggesting that these agents do not kill cells by the induction of visible chromosome breaks. An alternative mode of death, apoptosis, has been demonstrated to be the mechanism of MMC and *cis*-

platinum cytotoxicity for several types of non-FA cells (e.g., LOWE et al. 1993; REY et al. 1994; BORSELLINO et al. 1995; MIZUTANI et al. 1995). FA cells exposed to lethal amounts of DNA cross-linking agents typically arrest irreversibly in the first G_2 following exposure (e.g., KAISER et al. 1982; DEAN and FOX 1983; SEYSCHAB et al. 1995) rather than divide and missegregate damaged chromosomes. This behaviour is similar to that of A-T cells, which arrest in G_2 and then die from apoptosis following X-irradiation (MEYN et al. 1994, submitted).

These observations raise the possibility that apoptosis plays a role in the killing of FA cells by DNA cross-linking agents. However, investigations of mutagen-induced apoptosis in FA cells have produced mixed results. ROSSELLI et al. (1995) found that FA(C) and FA(D) lymphocytes expressed increased spontaneous apoptosis, but the induction of apoptosis by X-irradiation was *diminished* in these FA cells. As to the induction of apoptosis in FA cells by DNA cross-linking agents, REY et al. (1994) reported that both control and FA cells underwent the same degree of apoptosis following exposure to highly toxic concentrations of MMC (10 μg/ml). However, we have found that exposure to low doses of either MMC (0.6 μg/ml) or DEB (4 ng/ml) induces proportionately much more apoptosis in FA(A), FA(C), and FA(D) lymphoblasts than in controls (ALLEN and MEYN 1995). A more definitive evaluation of the possible role of apoptosis in FA may come from ongoing studies of FA knockout mice as well as more extensive studies of FA cells in culture.

2.4.10 Lymphokines and FA

Several recent observations provide evidence that lymphokine abnormalities may play a role in the FA phenotype. ROSSELLI et al. (1992) found that addition of interleukin-6 (IL-6) to culture media corrected the sensitivity of FA(A) and FA(B) lymphoblasts to the clastogenic and cytotoxic effects of MMC, while addition of IL-1, IL-2, or IL-3 had no effect. The correction by IL-6 was specific for FA cells and could be inhibited by the addition of anti-IL-6 antibody. The FA lymphoblasts used in this study were found to have decreased secretion of IL-6 in culture and were insensitive to the induction of IL-6 by exposure to tumor necrosis factor alpha (TNFα). Production of IL-6 and the cytokine granulocyte–macrophage colony-stimulating factor (GM-CSF) have been found to be consistently decreased in both long-term bone marrow cultures and peripheral blood mononuclear cells from FA homozygotes (BAGNARA et al. 1993; STARK et al. 1993; WUNDER et al. 1993), although BAGBY et al. (1993) found mixed results in FA fibroblasts.

Tumor necrosis factor alpha also may be dysregulated in FA cells. Reporting the results of a survey of cytokine production in lymphoblasts, ROSSELLI et al. (1994) found up to eightfold increases over controls in the amounts of TNFα found in conditioned media from lymphoblasts representing FA complementation groups A, B, C, and D. These increases were not due to detectable differences in the genomic structure of the *TNFα* gene or TNFα mRNA expression. They also found that 36 FA homozygotes and 21 FA heterozygote parents all had serum levels of TNFα that were higher than 14 controls, confirming the previous observation of

SCHULTZ and SHAHIDI (1993). The relevance of overproduction of this cytokine to the rest of the FA phenotype was suggested by the demonstration that addition of anti-TNFα antibodies to the media of FA lymphoblasts partially corrected their sensitivity to the cytotoxic and clastogenic effects of MMC (ROSSELLI et al. 1994).

2.5 FA Genes

Unlike A-T, the multiple complementation groups of FA represent at least three separate genes: *FACC*, the gene for FA(C), at 9q22.3 (STRATHDEE et al. 1992a), the locus for FA(A) at 16q24.3 (PRONK et al. 1995), and the FA(D) locus on 3p (WHITNEY et al. 1995). Somatic cell hybrid studies indicate that there are at least five complementation groups (JOENJE et al. 1995), raising the possibility of five separate FA genes.

FACC, the only FA gene isolated to date, is expressed ubiquitously, although expression is particularly high in mesenchymal tissues and bone during embryonic development in the mouse (STRATHDEE et al. 1992b; WEVRICK et al. 1993). It codes for a 558-amino acid protein with no known homologies (STRATHDEE et al. 1992b), produces three different mRNAs due to alternate polyadenylation sites and has two alternative 5′ exons which do not alter the coding sequence (CHEN et al. 1995; SAVOIA et al. 1995). The most 5′ exon is used for constitutive expression, while the other appears to be preferentially expressed following DNA damage (CHEN et al. 1995). The normal function of *FACC* is not clear. Somewhat surprisingly, immunohistochemical analyses indicate that the FACC protein is part of a complex of cytosolic proteins and does not enter the nucleus, even in response to DNA damage (YAMASHITA et al. 1994; YOUSSOUFIAN 1994; YOUSSOUFIAN et al. 1995).

Mouse knockouts for *FACC* have been created using deletions of exon 8 or exon 9 (CHEN et al. 1995; GROMPE et al. 1995). Fibroblasts from mice homozygous for either of the *FACC* deletions express sensitivity to bifunctional cross-linking agents, and the female homozygotes have markedly reduced fertility (CHEN et al. 1995; GROMPE et al. 1995). However, the mice do not have widespread malformations, nor have they yet developed pancytopenia or leukemia (CHEN et al. 1995; GROMPE et al. 1995).

2.5.1 FA Genes and Phenotypic Variability

Variability of phenotypic expression is common, as indicated by a recent analysis of the IFAR data, in which 39% of 328 FA homozygotes were found to have had both hematological problems and malformations, 24% had only malformations, and the rest only hematological abnormalities (AUERBACH et al. 1989). The existence of multiple FA genes and mutations may contribute to the clinical variability of FA. For example, the *IVS4* mutation of the *FACC* gene is associated with an unusually early age at onset of pancytopenia and more than the usual frequency of congenital malformations (VERLANDER et al. 1994). However, some phenotypic variability may be independent of allele differences in FA genes, as indicated by lack of concordance for congenital malformations between monozygotic twins and siblings (GLANZ and FRASER 1982; ADLER-BRECHER et al. 1992; GIAMPIETRO et al. 1993, 1994).

2.6 Underlying Defect(s) in FA

In 1969, COHEN and LEVY wrote that "a potentially unifying theory capable of explaining the diverse clinical, cellular and molecular manifestations of FA has yet to emerge." Twenty-seven years later, we still do not understand the molecular pathology that underlines the pleiotropic phenotype of FA. Several different hypotheses have been put forth. One general explanation assumes that an underlying deficiency in the handling of certain types of spontaneous and induced DNA damage is responsible for both the in vivo and in vitro abnormalities seen in FA. In this model, an inability to correctly repair DNA damage results in the accumulation of chromosome breaks and other forms of genetic damage. This damage to the genome can lead in turn to: (a) death during embryogenesis of cells responsible for normal limb and other organ formation; (b) postnatal death of bone marrow stem cells; and (c) the inactivation of tumor suppressor genes and activation of oncogenes in somatic tissues. This combination of increased prenatal and postnatal cell death and a genetic instability phenotype leads to the congenital malformations, bone marrow failure, and cancer risk seen in FA homozygotes. The most straightforward versions of this model postulate that FA mutations occur in genes coding for enzymes directly involved in the recognition, processing, and repair of DNA cross-links (e.g., LATT et al. 1975; SASAKI 1975; FUJIWARA and TATSUMI 1977; ALTER 1994c).

Another model emphasizes the sensitivity of FA cells to oxygen and other ROS, suggesting that the intrinsic defects in FA are in genes involved in cellular defenses against reactive oxygen radicals. In one form of this hypothesis, a pro-oxidant intracellular environment secondarily impairs the enzymatic machinery for interstrand DNA cross-link repair. This, in turn, leads to bone marrow failure, mutagen sensitivity, genetic instability, and cancer (SCHINDLER and HOEHN 1988; JOENJE and GILLE 1989; DEGAN et al. 1995). Alternatively, the bone marrow failure is thought to be a direct toxic effect of ROS on stem cells, rather than the gradual accumulation of oxidative damage to DNA (JOENJE and GILLE 1989).

ROSSINI et al. (1992, 1994) have proposed a third hypothesis for FA, one that developed out of their studies of lymphokine abnormalities in FA. They suggest that FA genes control "the expression of a network of genes involved in the development of the hematopoietic system," including *IL-6* and *TNFα*. In this model, mutations in FA genes cause perturbations in this network which, in an unspecified manner, lead to bone marrow failure, alterations in the oxidative state of cells, and an inability to properly repair DNA damage.

We recently suggested a fourth possibility, that dysfunctional apoptosis is involved in several aspects of the FA phenotype (ALLEN and MEYN 1995). The radial ray defects as well as other congenital anomalies seen in FA homozygotes could result from inappropriate apoptosis during embryonic morphogenesis, increased spontaneous apoptosis in bone marrow stem cells could account for the pancytopenia and a low threshold for DNA damage-induced apoptosis could be responsible for the sensitivity of FA cells to cross-linking agents. However, as discussed in Sect. 2.4, the possible involvement of apoptosis in the FA phenotype has yet to be thoroughly examined, even though several experimental observations are suggestive.

How might defects in FA genes lead to dysfunctional apoptosis? As discussed previously, FA cells do not appear to have the global defects in DNA damage responses that are seen in A-T. An alternative explanation as to how apoptosis might be involved in the FA phenotype comes from the studies of ROSSINI et al. (1992, 1994), who documented abnormalities of IL-6 and TNFα in FA. TNFα is a potent negative regulator of hematopoietic precursor cells in vivo that appears to suppress the growth of bone marrow stem cells through the activation of FAS-mediated apoptosis (NAGAFUJI et al. 1995; SELLERI et al. 1995; ZHANG et al. 1995). In addition to its role in life/death decisions in the bone marrow, TNFα can induce apoptosis in non-hematopoietic cells (reviewed in LARRICK and WRIGHT 1990; SCHWARZ et al. 1995). TNFα may also be involved in modulating apoptosis during embryogenesis (reviewed in WRIDE and SANDERS 1995). TNFα is expressed in early embryos (ROTHSTEIN et al. 1992; KOHCHI et al. 1994; WRIDE et al. 1994), and anti-TNFα antibodies localize at sites within the embryo that are undergoing apoptosis (WRIDE et al. 1994). As noted by WRIDE and SANDERS (1995), thalidomide is a potent inducer of limb malformations that specifically alters the stability of TNFα mRNA (MOREIRA et al. 1993), raising the possibility that TNFα normally plays a role in limb development and that the drug exerts its teratogenic effects by disrupting TNFα expression at a critical time during embryogenesis.

TNFα expression affects the oxidative state of cells. Intracellular expression of recombinant TNFα increases oxidative damage to DNA and other macromolecules (ZIMMERMAN et al. 1989), and treatment of cells in culture with exogenous TNFα causes the accumulation of free radicals within the cell (reviewed in LARRICK and WRIGHT 1990). The ability of TNFα to induce apoptosis in cells grown in culture can be prevented by growth in anaerobic conditions (MATTHEWS et al. 1987) or treatment with antioxidants (ZIMMERMAN et al. 1989; LARRICK and WRIGHT 1990), suggesting that the triggering of apoptosis by TNFα is mediated, at least in part, by ROS.

In contrast to the apoptosis-promoting effects of TNFα, IL-6 appears to act as an anti-apoptosis factor, as indicated by its ability to prevent p53-mediated apoptosis in myeloid leukemia cells (YONISH-ROUACH et al. 1991). Administration of IL-6 to mice protects their bone marrow from the lethal effects of ionizing radiation (PATCHEN et al. 1991), and treatment of renal carcinoma or prostate carcinoma cells with monoclonal antibodies against IL-6 synergistically potentiates the cytotoxic effects of *cis*-platinum and MMC (BORSELLINO et al. 1995; MIZUTANI et al. 1995). These results suggest that a normal function of IL-6 is to inhibit DNA damage-induced apoptosis. This is in contrast to the observation that TNFα can act synergistically to potentiate apoptosis induced by ionizing and UV radiation (e.g., HALLAHAN et al. 1990; SCHWARZ et al. 1995).

Taken together, these observations about TNFα and IL-6 suggest that much of the FA phenotype may be due to dysregulation of these, and perhaps other, lymphokines. It has been suggested that TNFα plays a role in bone marrow suppression in sporadic aplastic anemia through the induction of FAS-mediated apoptosis (MACIEJEWSKI et al. 1995), and apoptosis is thought to play a role in another hereditary anemia, Diamond-Blackfan syndrome (PERDAHL et al. 1994). Overexpression of TNFα could be responsible for the ongoing depletion of bone marrow stem cells that is thought to underlie the development of aplastic anemia in FA homozygotes. Dysregulation of TNFα could lead to inappropriate apoptosis during embryonic development, accounting for the congenital malformations seen in FA homozygotes. In addition, high levels of TNFα could be responsible for the increased spontaneous frequency of apoptosis seen in FA cells grown in culture. The "prooxidative" state of FA cells, their increased baseline oxidative DNA damage, and the poor growth of FA cells in the presence of oxygen may also be the result of inappropriately high levels of TNFα. Increases in the amount of spontaneous and induced oxidative DNA damage in FA cells could in turn be responsible for their chromosome instability.

TNFα and IL-6 exert opposing effects on the cellular decision to undergo apoptosis following DNA damage. The increased levels of TNFα and decreased levels of IL-6 seen in FA would predict that, for a given level of DNA damage, FA cells would be more likely to undergo apoptosis than controls, as we have documented for FA cells grown in culture (ALLEN and MEYN 1995). Proposing that the sensitivity of FA cells to the cytotoxic effects of DNA cross-linking agents is the result of a low threshold for triggering apoptosis resolves several issues. It explains why investigators have not found a consistent defect in DNA cross-link repair in the various FA complementation groups, why FA cells repair damaged viral DNA to the same extent as normal cells, and how FACC, an exclusively cytoplasmic protein, affects the survival of cells that have sustained DNA damage.

Based on map position, sequence information, and gene expression, the *TNFα* and *IL-6* genes are not FA genes (e.g., ROSSELLI et al. 1994). If they are playing a pivotal role in the FA phenotype, then it may be, as suggested by ROSSELLI et al. (1994), because FA genes disturb the tightly regulated expression of these critical members of a signal transduction network that regulates cell growth and death (BAKER and REDDY 1996). How multiple FA genes might affect lymphokine regulation so as to have similar phenotype is not clear. They may form a functional complex together, or they may work in the same pathway. There is much work to be done before this hypothesis can be validated and these issues resolved.

2.6.1 Genetic Instability and Cancer in FA Homozygotes

Whatever the nature of the underlying defects in FA, the spontaneous and induced genetic instability seen in this condition is presumably behind much of the cancer risk. Despite their genetic instability, FA cells are relatively hypomutable for point mutations (PAPADOPOULO et al. 1990a; LAQUERBE et al. 1995), suggesting that deletions and other rearrangements play key roles in the accumulation of genetic damage necessary for the development of cancer in this syndrome. It has been suggested that the clonal abnormalities seen in the bone marrow of FA homozygotes predispose to the development of AML, since the leukemia risk is higher in those FA homozygotes with karyotypic abnormalities of their bone marrow stem cells (BUTTURINI et al. 1994). However, clonal abnormalities in FA bone marrow wax and wane (ALTER 1994a), and an etiologic link between the occurrence of karyotypically abnormal bone marrow stem cells and the development of leukemia has yet to be established.

2.7 BS Phenotype

Over 40 years ago, BLOOM (1954) described "congenital telangiectatic erythema resembling lupus erythematous in dwarfs." Nine years later, SZALAY (1963) reported familial clustering of this condition, which later was shown to be the result of autosomal recessive inheritance (GERMAN 1969). BS homozygotes typically survive into their early 20s (GERMAN 1995), despite the clinical and laboratory abnormalities summarized below. Additional phenotypic detail is available in recent reviews by GERMAN (1993, 1995).

2.7.1 Growth Deficiency

Proportional dwarfism is a constant finding in BS homozygotes, whose average size and weight at term birth are at the 50th percentile for a 32-week premature infant (GERMAN 1995). They remain well below the third percentile in all growth parameters throughout their lives. Slow growth is also a well-known attribute of BS cells grown in culture.

2.7.2 Immunodeficiency

LANDAU et al., (1996) first documented low levels of circulating Igs in BS homozygotes, and most individuals have defective delayed hypersensitivity as well (reviewed in WEEMAES et al. 1991). These generalized immunodeficiencies contribute to the serious bacterial sinopulmonary infections that occur in these children, making chronic respiratory failure the second leading cause of death (GERMAN 1995).

2.7.3 Malignancy

Bloom syndrome was the first chromosome instability syndrome in which a link was recognized between genetic instability and cancer risk (GERMAN et al. 1965). BS is a much less common condition than A-T and FA, and there is only one large ongoing study of BS homozygotes, the Bloom's Syndrome Registry established in the early 1960s by James German. The following discussion of malignancy in BS is based on data contained in a recent analysis of the first 165 BS homozygotes entered into the registry (GERMAN 1993).

As expected for a cancer predisposition syndrome, there is a high risk of cancer; tumors occur early in life and multiple primary neoplasms are frequent. Of the individuals in the Bloom's Syndrome Registry report of 1993, 37% had developed malignant tumors, with a mean age at onset of 24.4 years; 28% of the BS homozygotes with cancer developed more than one primary tumor, and five of 60 had three or more malignancies (GERMAN 1993). The result is that cancer is the leading cause of death in BS homozygotes, accounting for 80% of deaths seen in the Registry (GERMAN 1995).

As shown in Fig. 4, lymphoreticular malignancies are the most common form of cancer in BS homozygotes, with leukemias and lymphomas occuring in roughly equal proportion. Together, they accounted for 45% of the malignant tumors reported in the 1993 survey (GERMAN 1993). As with A-T, lymphomas and leukemias are most common in younger patients, although the peak ages for lymphoreticular malignancies in BS are 11–25 years rather than the 6–10 years seen in A-T (compare Figs. 1 and 4). A variety of common solid tumors account for the remaining malignancies in BS homozygotes (GERMAN 1993), the most frequent carcinomas being colon, skin, breast, and cervix. Less common solid tumors include carcinomas of the tongue, larynx, lung, and esophagus as well as Wilms tumor, osteosarcoma, retinoblastoma, and medulloblastoma (GERMAN 1993; GIBBONS et al. 1995).

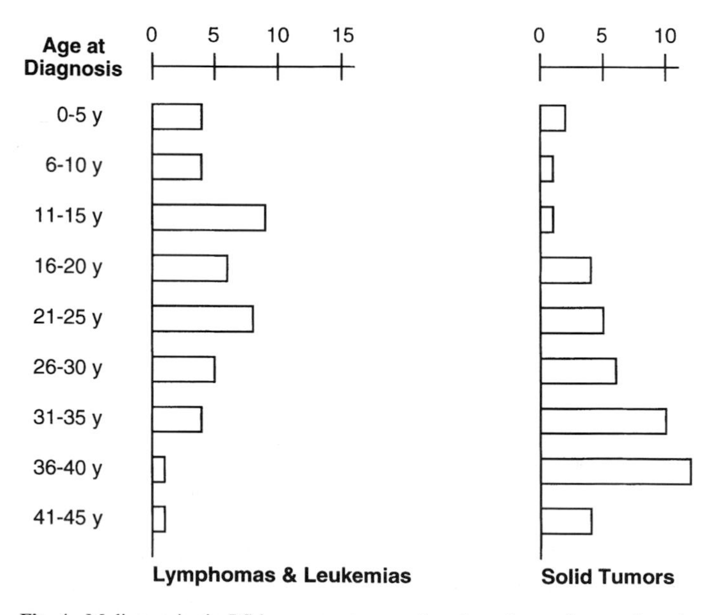

Fig. 4. Malignancies in BS homozygotes as a function of age of onset (based on data from the Bloom's Syndrome Registry report of 1993; GERMAN 1993)

Non-surgical treatment of cancer in BS is made difficult by the apparent sensitivity of some, but not all, BS homozygotes to radiation therapy and/or chemotherapy. Several case reports have documented severe and prolonged marrow suppression, esophageal strictures and other adverse effects following normal courses of radiation or chemotherapy (e.g., PASSARGE 1991; OTO et al. 1992). Unlike A-T homozygotes, however, an in vivo sensitivity to DNA-damaging agents does not appear to be universal in BS since many BS homozygotes have tolerated normal therapeutic regimens without incident (PASSARGE 1991).

2.7.4 Other Clinical Abnormalities

Bloom Syndrome homozygotes express a characteristic sun-sensitive erythematous facial rash that usually develops in the first summer of life (BLOOM 1954; GERMAN 1993). They frequently develop hyper- and hypopigmented macular skin lesions, which may represent dermatologic manifestations of increased somatic genetic recombination. Like A-T, gonadal abnormalities are a feature of the syndrome. Men with BS have azoospermia, and female homozygotes tend to have irregular menses as well as early menopause. However, some women with BS have borne children (GERMAN 1993). Diabetes has been reported in older BS homozygotes (MORI et al. 1990; GERMAN 1993). BS homozygotes typically express average or near-average intelligence, although learning disabilities are frequent and there are occasional BS homozygotes with moderate mental retardation (GERMAN 1995). Like A-T homozygotes, the personality of BS homozygotes tends towards an innate optimism. Major congenital malformations are not seen in BS homozygotes, although they have a distinctive appearance and high-pitched voice due to their small, dolichocephalic skulls, malar hypoplasia, protuberant ears, and small mandibles (GERMAN 1995).

2.7.5 Spontaneous Genetic Instability

Genetic instability is a hallmark of the BS phenotype. Beginning with GERMAN'S initial observations of an increased frequency of chromosome rearrangements, (GERMAN 1964; GERMAN et al. 1965), a variety of spontaneous chromosomal aberrations have been observed in cells from BS homozygotes, including SCEs, quadriradials, telomere associations, chromosome breaks, chromosome gaps, and micronuclei (reviewed in KUHN and THERMAN 1986; COHEN and LEVY 1989). SCEs and quadriradials (chromosome aberrations that involve recombinational exchanges between homologous chromosomes) are especially frequent in BS homozygotes (e.g., CHAGANTI et al. 1974). Multiple cell lineages express chromosomal instability in BS, including lymphocytes, lymphoblasts, keratinocytes and fibroblasts (COHEN and LEVY 1989).

Spontaneous genetic instability in BS homozygotes also has been documented by molecular and immunologic analyses. Studies of the glycophorin locus in erythrocytes found 50- to 100-fold elevations in frequencies of conversion of *M* and *N* alleles to hemizygosity and homozygosity in cells from BS homozygotes (KYOIZUMI et al. 1989; Langlois et al. 1989). BS T lymphocytes have significant elevations in their spontaneous frequencies of *HPRT* mutations (Evans et al. 1983), and tandem repeats rearrange more frequently in BS lymphoblasts than controls (Groden and German 1992). Increased frequencies of spontaneous mitotic chromosomal recombination have been documented in BS T lymphocytes in vivo (Groden et al. 1990) and in B lymphoblasts and fibroblasts in culture (Brainard et al. 1991; Kusunoki et al. 1994). Extrachromosomal recombination also is elevated (Bubley and Schnipper 1987), suggesting that generalized hyperrecombination is an intrinsic feature of BS. Finally, it should be noted that the *BLM* locus itself is a target for interchromosomal recombination in BS cells, with intralocus rearrangements restoring a wild-type gene and normal genetic stability (Ellis et al. 1995b).

2.7.6 Abnormal Responses to DNA Damage

Bloom Syndrome homozygotes are not as sensitive to DNA-damaging agents as A-T homozygotes or FA homozygotes. However, there is in vivo and in vitro evidence for decreased resistance to the killing effects of these agents. As indicated above, BS homozygotes can be more sensitive to therapeutic radiation and chemotherapy than normal individuals. Cells from BS homozygotes exhibit sensitivity in culture as well. Most BS cells are moderately sensitive to the killing effects of ethyl methanesulfonate (EMS) (Arlett and Harcourt 1978; Krepinsky et al. 1989), *N*-ethyl-*N*-nitrosourea (Kurihara et al. 1987a), and 5-fluorouracil (Lonn et al. 1990). Many BS cells are sensitive to X-rays (Arlett and Harcourt 1980; Weichselbaum et al. 1980; Hall et al. 1986; Costa and Thacker 1993) and UV light (Krepinsky et al. 1980; Ishizaki et al. 1981), although the degree of sensitivity is variable, with some BS cells expressing essentially normal resistance to these agents and to double-strand DNA breaks induced by restriction enzyme treatment (Costa and Thacker 1993).

Another part of BS phenotype is a low threshold for the induction of chromosomal aberrations by many DNA-damaging agents. Following treatment with UV light (Krepinsky et al. 1980; Kurihara et al. 1987b), ionizing radiation (Evans et al. 1978; Kuhn 1980; Aurias et al. 1985a), EMS (Krepinsky et al. 1979), or bromodeoxyuridine (BrDU) (Heartlein et al. 1987), cells from BS homozygotes typically express significantly more chromosome breaks, SCEs and/or quadriradials than controls. In contrast, the cross-linking agents MMC and DEB do not induce excess SCEs in BS cells (Shiraishi and Sandberg 1978; Auerbach and Wolman 1979), even though BS cells are sensitive to the killing effects of MMC (Ishizaki et al. 1981).

Although abnormalities in DNA repair enzymes have been documented in BS cells (see below), BS cells do not have major functional defects in the several DNA repair pathways that have been examined to date (German and Schonberg 1980; Cohen and Levy 1989; Evans and Bohr 1994). Damage-induced cell cycle checkpoints have not been extensively studied in BS. However, the G_1/S checkpoint is triggered normally after UV irradiation in BS fibroblasts (Van Laar et al. 1994), despite impaired induction of p53 following irradiation in some BS cell lines (Lu and LANE 1993). The p53-independent, X-ray-induced, S phase checkpoint also appears to be intact (YOUNG and PAINTER 1989).

2.8 *BLM* Gene

Apparent abnormalities in the structure, activity, or expression of a variety of enzymes involved in DNA metabolism have been found in BS cells, including DNA ligase I (CHAN et al. 1987; WILLIS and LINDAHL 1987), topoisomerase II (HEARTLEIN et al. 1987) thymidylate synthetase (SHIRAISHI et al. 1989), uracil-DNA glycosylase (VOLLBERG et al. 1987; SEAL et al. 1988), N-methylpurine DNA glycosylase (DEHAZYA and SIROVER 1986), O^6-methylguanine methyltransferase (KIM et al. 1986), and SOD (NICOTERA et al. 1989). However, the recent positional cloning of the gene responsible for BS indicates that these abnormalities are, at best, epiphenomena.

Unlike A-T and FA, a single complementation group defines BS (WEKSBERG et al. 1988). Four years ago, MCDANIEL and SHULTZ (1992) used microcell-mediated chromosome transfer to demonstrate that chromosome 15 complements the high spontaneous SCE phenotype of BS cells in culture, suggesting that the single locus for BS lay on chromosome 15. This observation facilitated the work of German's group, who subsequently linked the BS locus to chromosome 15q26.1 markers (GERMAN et al. 1994), using homozygosity mapping in inbred families (LANDER and BOTSTEIN 1987). They subsequently isolated a gene from 15q26.1 that was mutated in multiple BS homozygotes (ELLIS et al. 1995a). The gene, termed *BLM,* codes for a 4.4-kb RNA transcript. It is homologous to the *S. cerevisae SGS1* and *E. coli recQ* genes, members of the DExH family of DNA/RNA helicases, which also include the mammalian DNA repair genes *XPB*, *XPD*, and *ERCC6* (ELLIS et al. 1995a).

Mutations in *SGS1*, the most closely related gene to *BLM* , share phenotypic similarities with BS homozygotes. The phenotype of *sgs1* null mutants includes slow growth, poor sporulation, increased mitotic nondisjunction, missegregation at meiosis, and high frequencies of both inter- and intra-chromosomal homologous mitotic recombination (GANGLOFF et al. 1994; WATTS et al. 1995). Sgs1p normally interacts in vivo with the topo II and Top3p topoisomerases and *sgs1* mutants suppress the phenotype of *top3* mutants (GANGLOFF et al. 1994; WATTS et al. 1995), suggesting that the Sgs1p helicase functions together with these topoisomerases in an in vivo complex. The BLM protein may interact with the human homologues to the yeast topo II and Top3p topoisomereases (TSAI-PFLUGFELDER et al. 1988; AUSTIN et al. 1993; FRITZ et al. 1996) since the topoisomerase activity of BS cells is abnormally low following exposure to BrDU (HEARTLEIN et al. 1987).

The sequence and phenotypic similarities between *SGS1* and *BLM* described above suggest that these two genes normally perform similar functions. It is not known at the present time what these functions might be. However, as suggested by GANGLOFF et al. (1994) for Sgs1p, the putative BLM helicase may help resolve intertwined strands at replication forks during DNA replication. This hypothesis is supported by observations of DNA synthesis abnormalities in BS cells (Hand and German 1975; Ockey and Saffhil 1986; Lonn et al. 1990). The BLM helicase also may facilitate separation of chromosomes at mitosis and meiosis, as well as unwind strands so as to allow transcription and repair complexes access to damaged DNA. Finally, as discussed below, BLM could help maintain genetic stability by functioning as an anti-recombination protein.

2.9 BS Phenotype as a Manifestation of Somatic Genetic Instability

German (1993) has speculated that the growth retardation seen in BS homozygotes is the result of their tissues containing abnormally small numbers of cells. This could be the result of a defective BLM helicase interfering with DNA synthesis and/or chromosome separation, causing a prolonged cell cycle and slow growth both in vivo and in vitro. Alternatively, German (1993) has proposed that a model for BS in which growth retardation and other phenotypic abnormalities of BS homozygotes are the result of the deleterious effects of genetic instability. In this somatic mutation model for BS, growth retardation is secondary to lethal mutations and/or uniparental disomies resulting from somatic recombination, immune deficiencies are caused by disturbances in immune gene recombination, male sterility is due to disruption of crossing-over during meiosis, altered pigmentation, and other minor developmental anomalies are secondary to increased genetic damage, and cancer is the end result of a generalized increase in genetic instability.

2.9.1 How Defects in the BLM Helicase Might Promote Genetic Instability

Defects in the BLM helicase could lead to genetic instability via one or more of several different routes:

1. Inability to resolve intertwined strands during replication may lead to DNA breaks, which are highly recombinogenic.
2. The BLM helicase may be involved in decatenating sister chromatids at mitosis and homologous chromosomes at meiosis. If so, *BLM* mutations could lead to an impaired ability to separate homologous DNA strands at cell division, thereby causing increased aneuploidy as well as chromosome breakage.
3. Wild-type BLM protein may directly inhibit genetic recombination, either by disrupting nascent heteroduplex formation during the strand invasion phase of recombination or by promoting the parental resolution of Holliday structures. In this scenario, absence of a functional BLM helicase would allow more nascent recombinational events to proceed to completion, resulting in an increased frequency of SCEs, quadriradials and other homologous recombinational events.
4. Finally, like the RAD3 DNA helicase of *S. cerevisiae* (Sung et al. 1987), BLM protein normally may unwind DNA so as to facilitate access to damaged DNA by repair complexes. Loss of BLM function in BS homozygotes would limit the access of repair complexes to sites of DNA damage, and the resultant increase in

unrepaired DNA lesions would contribute to the mutagen sensitivity, chromosome aberrations and genetic instability seen in these individuals.

2.9.2 Genetic Instability and Cancer in BS Homozygotes

The lack of a DNA helicase in the somatic cells of BS homozygotes presumably increases their frequency of spontaneous and induced chromosome aberrations, mitotic recombination, and loss of heterozygosity. The result is a loss of control of genetic integrity and an unusually rapid accumulation of the multiple genetic changes in tumor suppressor genes and oncogenes necessary to give rise to tumors. The wide variety of common tumors that occur in young BS homozygotes provides strong support for the hypothesis that acquisition of a genetic instability phenotype is a key step in oncogenesis. The immune deficiencies seen in BS homozygotes, as well as their high risk of leukemia and lymphoma, suggest that the BLM helicase may play a role in immune gene rearrangement as well as support the idea that the initial production of strand breaks and other DNA damage is a rate-limiting step in tumor development, even in genetically unstable cells.

3 What Is the Effect of Germ Line Heterozygote Mutations on Cancer Risk and Genetic Instability?

As detailed in the previous sections, a great deal of progress has been made in the last few years towards understanding the molecular pathology of the chromosome instability syndromes. These advances may help us address an important practical consideration, i.e., the extent to which mutations in the chromosome instability syndrome genes contribute to the development of cancer in the general population.

In evaluating the potential impact of the chromosome instability syndrome genes on the overall human cancer risk it is useful to consider what effects germ line mutations in these genes might have on the incidence of familial and sporadic cancers. As summarized previously, people who are homozygous for germ line mutations in chromosome instability syndrome genes are at high risk for a variety of tumors. However, these individuals represent a very small fraction of the total number of patients with cancer (see below). On the other hand, the heterozygote carriers are much more common. If they are at increased risk to develop cancer, then they might account for a significant proportion of cancer patients. The following section reviews published data concerning cancer risk and other phenotypic abnormalities in chromosome instability syndrome heterozygotes and discusses their likely contribution to the total burden of familial and sporadic cancers.

3.1 Are Chromosome Instability Syndrome Heterozygote Carriers at Risk for Cancer?

Individuals who carry germ line mutations in certain genes that affect genetic stability are known to have a high cancer risk. For example, both humans and mice who are heterozygotes for mutations in *p53* have a high incidence of lymphoma and other cancers (Birch et al. 1994; Harvey et al. 1995), and patients with germ line mutations in the *hMLH1* and *hMSH2* mismatch repair genes are at increased risk for colon and endometrial cancers as well as other solid tumors (reviewed in Lynch and Lynch 1994). Given these examples, one might expect that heterozygote carriers of mutations in the *ATM*, *BLM*, or FA genes would be at increased risk to develop cancer.

As summarized below, the published literature supports the conclusion that *ATM* heterozygotes face an approximately fourfold higher risk of breast cancer than normal individuals. They may also have an elevated risk for other tumors. In contrast, existing evidence for any increased cancer risk in FA heterozygotes is slim, and almost non-existent for carriers of *BLM* mutations.

3.1.1 Ataxia Telangiectasia

Direct support for an increased cancer risk in A-T heterozygotes comes from studies of the relatives of A-T homozygotes. An increased incidence of cancer amongst relatives of A-T homozygotes was first noticed by Reed et al. (1966). In 1976, Swift et al. reported a study of 27 A-T families in which relatives of A-T patients suffered an excess of deaths from cancer over that expected in the general population (59 deaths observed vs. 42.6 expected, $p < 0.02$). A subsequent retrospective study of 110 additional families confirmed and expanded the original findings (Swift et al. 1987). This was followed by a 1991 prospective study of 1420 individuals in 161 A-T families (Swift et al. 1991) which found that the rate ratios of all cancers in blood relatives when compared to spouse controls were 2.5 for men and 2.1 for women. The ratios for obligate heterozygotes were higher (3.9 for males and 2.7 for females). Relative risk estimates for heterozygotes were calculated to be 3.8 for males and 3.5 for females ($p < 0.005$).

In Swift's studies, the increased occurrence of cancer in blood relatives of A-T patients was particularly striking for breast cancer, although there also were excess numbers of lung, ovarian, pancreatic, bladder, and stomach cancers (Swift et al. 1987, 1991). Swift's conclusions that A-T heterozygotes have an increased relative risk for breast cancer have been supported by two smaller studies. In a survey of 67 A-T families in Great Britain by Pippard et al. (1988), researchers found an excess of breast cancer in the parents of A-T homozygotes, as did a study of eight Norwegian families by Borresen et al. (1990). The observed increases in breast cancers amongst blood relatives of A-T homozygotes were due primarily to an excess of cancers in women under 55 and, in the largest survey (Swift et al. 1991), 25% of female blood relatives with breast cancer had two independent cancers. Although these data sets are not large, the excess numbers of cancers, shift in the age of onset towards younger women, and elevated frequency of multiple primary tumors are what might be expected for individuals with a hereditary predisposition to develop breast cancer.

In a recent meta-analysis, EASTON (1994a) reviewed the published family studies and found that the combined data supported estimates of a relative risk of 3.9 (95% confidence interval of 2.1–7.2) for developing breast cancer in female A-T heterozygotes (see Table 1). EASTON (1994a) concludes that these studies provide compelling evidence for an increased risk of breast cancer in A-T heterozygotes. The case for an increased risk of breast cancer in A-T heterozygotes is further supported by unpublished work of SWIFT (1996) in which polymorphic markers linked to the *ATM* locus were used to identify heterozygotes in A-T families. This method of analysis avoids some of the ascertainment bias problems inherent in previous studies. Of the 27 A-T family members with breast cancer that have been analyzed to date, 21 were found to be A-T heterozygotes.

These same family studies do not present as strong a case for the proposition that A-T heterozygotes face an increased risk of developing cancers other than breast carcinoma. In EASTON's meta-analysis (1994a), the combined data sets gave a relative risk of 1.9 (95% confidence interval of 1.5–2.5) for all cancers in A-T heterozygotes. However, as EASTON points out, the four published surveys are inconsistent regarding the overall risk of cancer. SWIFT's retrospective and prospective studies of American A-T families predict 2.6–3.6 increases in the relative risk of developing cancer in A-T heterozygotes, while the smaller British and Norwegian surveys estimate that the overall risk of developing cancer is essentially normal in these individuals. Resolution of the discrepancy between these studies may require both addition of new families to the study populations as well as identification of A-T heterozygotes within these families by molecular analysis.

The DSN model for A-T summarized above presumes that p53 is downstream of ATM in one or more pathways that mediate cellular responses to DNA damage. Given the assumption of this and other models that ATM and p53 are in the same signal transduction network (ENOCH and NORBURY 1995; KASTAN 1995; MEYN 1995), it is useful to compare the spectrum of tumors seen in heterozygotes and homozygotes for *ATM* and *p53* mutations.

Table 1. Estimated relative risks of cancer in A-T heterozygotes (modified from EASTON 1994a)

Study	Breast cancers			Other cancers		
	Observed/expected (A-T blood relatives)	Relative risk to A-T heterozygotes Mean	Range	Observed/expected (A-T blood relatives)	Relative risk to A-T heterozygotes Mean	Range
SWIFT et al. (1987)	27/20.5	6.8	2.0–22.6	111/103	2.6	1.4–4.7
SWIFT et al. (1991)	23/7	5.1	1.5–16.9	68/29	3.5	2.0–6.1
BORRESEN et al. (1990)	6/2.4	3.9	1.26–12.1	14/20	0.77	0.32–1.85
EASTON (1994a)	2/0.17	1.3	0.33–5.2	25/24.5	1.0	0.59–1.7
	58/30	3.9	2.1–7.2	218/179	1.9	1.5–2.5

As can be seen in Table 2, both A-T homozygotes and *p53*-null mice have a predilection for immune system tumors. KASTAN et al. (1992) suggests that an abnormal response to DNA strand breaks may be responsible for the high incidence of lymphoid tumors in both cases. In the DSN model, this abnormal response would be the inability to trigger the ATM-dependent DSN and activate its cell cycle checkpoints. Breast cancer is common in both Li-Fraumeni family members and A-T blood relatives, yet is relatively infrequent in A-T homozygotes or *p53*-null mice. Death at an early age from other causes may reduce the apparent occurrence of breast cancer in homozygotes. Alternatively, these differences may reflect phenotypic differences between the heterozygote and homozygote states of these two genes.

The spectrum of tumors seen in A-T homozygotes is not identical to that in *p53*-null mice. For example, leukemias are the second most common tumor in A-T homozygotes but are not seen in *p53*-null mice, while sarcomas are a common tumor type in *p53*-null mice but are rarely seen in A-T homozygotes. The latter may reflect a functional difference between the two genes rather than a difference between the two species, since sarcomas are frequent in human *p53* heterozygotes (e.g., affected Li-Fraumeni family members).

3.1.1.1 Etiology of A-T Heterozygotes' Cancer Risk

Analogous to germ line mutations in the *Rb* gene predisposing to retinoblastoma, A-T heterozygotes could be at increased risk to develop cancer solely because they have only one functional *ATM* allele to inactive before they lose ATM-dependent damage responses and become genetically unstable. Alternatively, their increased cancer risk could be due, in part, to *ATM* mutations not being completely recessive. If cells from A-T heterozygotes are not phenotypically normal, then inherent genetic instability and mutagen sensitivity would directly facilitate the accumulation of somatic mutations necessary for tumor development. As detailed in Section 3.2.1, this possibility is supported by multiple studies demonstrating that *ATM*

Table 2. Tumor spectrum in *p53* and *ATM* homozygotes and heterozygotes

Tumor type	Genetic background (source)				
	Li-Fraumeni family members	$p53^{+/-}$ mice	A-T blood relatives	$p53^{-/-}$ mice	A-T homozygotes
Lymphoma (%)	–	31	5	55	56
Leukemia (%)	6	–	4	–	24
Sarcoma (%)	18	54	–	23	–
Breast carcinoma (%)	26	2	20	2	2
Brain tumor (%)	12	3	–	4	3
Other tumors (%)	38	10	71	18	15
Tumors analyzed (*n*)	231	60	138	90	107
Source	MALKIN (1993)	HARVEY et al. (1995)	SWIFT et al. (1987)	HARVEY et al. (1993)	SPECTOR et al. (1982)

heterozygotes have subtle abnormalities in their responses to DNA damage and their ability to maintain genomic stability.

3.1.2 Fanconi Anemia

In 1951 BAUMANN suggested that leukemia occurs unusually often in relatives of FA homozygotes, and, in a 1959 survey of 49 families with at least one FA homozygote, GARRIGA and CROSBY found leukemia in four "unaffected" relatives, a rate of one case of leukemia in a relative for every 12.2 FA homozygotes. They compared this rate to a previously reported rate of one case of leukemia in a blood relative for every 450 patients with leukemia and conducted that "the rate of leukemia in the Fanconi families seems to be astonishingly high," suggesting that this was due to "some hereditary fault." Subsequently, FANCONI (1967) wrote that "in the families with F.A., leukemia and other myelopathies are more frequent than in normal families." Two subsequent reviews of leukemia in FA homozygotes noted an apparent increased incidence of leukemia in blood relatives but offered no formal analysis thereof (DOSIK et al. 1970; SCHROEDER and KURTH 1971).

The putative link between FA heterozygote carriers and malignancy was formally investigated by SWIFT in 1971, who conducted a retrospective analysis of cancer in eight FA families. In this study, there were 27 cancer deaths observed in blood relatives of FA homozygotes, compared to the 17.4 expected ($p < 0.05$). There were excess deaths from leukemia (three observed, one expected), colon cancer (five observed, two expected), and stomach cancer (four observed, two expected). Cancer deaths were predominantly in the 50- to 74-year-old age group, however, and there was only one case of multiple primary tumors. SWIFT concluded that FA heterozygotes had threefold increase in their relative risk of cancer and that they might represent 5% of individuals with acute leukemia.

Nine years later, SWIFT et al. (1980) published an expanded study that included 1206 blood relatives from 25 FA families of diverse ethnic backgrounds. Forty-eight cancer deaths occurred in the blood relatives compared to 56.2 expected, and only three blood relatives had leukemia compared to 4.5 expected. There were no significant differences between cancer rates in blood relatives and controls, although there was a higher than expected number of tumors for four sites: gastric cancer (six observed, 4.9 expected), bladder cancer (seven observed, 4.8 expected), breast cancer (17 observed, 13.5 expected), and lung cancer (three observed, 0.6 expected). SWIFT et al. concluded that there was no overall excess of cancers in either FA blood relatives or obligate FA heterozygotes, although the question of increased risk of bladder, stomach, and breast cancer among FA heterozygotes was unresolved.

Subsequent to SWIFT et al.'s 1980 report, a smaller epidemiological study was conducted by POTTER et al. (1983), who surveyed 129 family members of nine index cases of FA. Seven malignancies were observed, as opposed to the 10.4 expected. Five of seven tumors occurred in individuals over age 60, and there were no leukemias. The authors concluded that there was no evidence for increased cancer risk in FA heterozygotes.

These studies suggest that, despite their abnormal responses to DNA-damaging agents (see Section 3.2.2), FA heterozygotes face little, if any, increase in their relative risks of developing cancer when compared to the population at large. However, these surveys cannot exclude the possibility that specific FA genes or mutant alleles of those genes might predispose to cancer in heterozygotes while the majority of FA mutations do not. The report of HILL (1976) is suggestive of this possibility. He described a remarkable inbred family in the Scottish Hebrides in which 13 of 21 immediate relatives of a FA homozygote had developed cancer, including three first cousins whose malignancies presented before age 10.

3.1.3 Bloom Syndrome

In an early report, GERMAN (1974) noted the occurence of five malignancies in 82 parents of BS homozygotes entered in the Bloom's Syndrome Registry. The tumors included colon, breast, and gastric carcinomas, as well as melanoma and myeloid dysplasia. Since then, formal analyses of cancer risk in BS heterozygotes have not been published, although one may be in the near future, using the families entered into the Bloom's Syndrome Registry (GERMAN 1995). GERMAN, who has had contact with the largest number of BS patients and families, states that he does not believe that BS heterozygotes have an increased cancer risk.

3.2 Phenotypic Abnormalities in Chromosome Instability Syndrome Heterozygotes

3.2.1 A-T Heterozygotes

3.2.1.1 Spontaneous Genetic Instability in A-T Heterozygotes

If A-T heterozygotes have an increased cancer risk because *ATM* mutations are not completely recessive, then one might expect increased genetic instability in cell from A-T heterozygotes. As summarized previously, spontaneous genetic instability in A-T homozygotes manifests itself in a variety of ways including increased chromosome aberrations, elevations in mitotic recombination rates (MEYN 1993), high mutation frequencies (BIGBEE et al. 1989; COLE and ARLETT 1994), increased aneuploidy (AGUILAR et al. 1968). Studies of these manifestations of genetic instability in A-T heterozygotes have yielded mixed results.

Several early studies found that lymphocytes and fibroblasts from A-T heterozygotes had modest, but statistically significant, elevations in their spontaneous frequencies of chromosomal aberrations (COHEN et al. 1975; OXFORD et al. 1975). In addition ROSIN et al. (1989), using micronuclei formation as a measure of chromosome breakage, documented that the average frequency of in vivo chromosome breaks in epithelial cells exfoliated from the oral mucosa of 26 A-T heterozygotes was threefold higher than that of 29 controls.

However, increased spontaneous chromosomal instability in A-T heterozygotes has not been confirmed by all investigators. Several studies found no obvious differences between A-T heterozygotes and controls (e.g., NATARAJAN et al. 1982; NAGASAWA et al. 1985). A recent examination by SCOTT et al. (1994) is typical. Studying lymphocytes from 27 heterozygotes, 14 homozygotes, and 66 age-matched controls, they found no significant differences between A-T heterozygotes and controls in the frequency of spontaneous chromosome aberrations.

Studies of individual genetic loci in small numbers of obligatory heterozygotes (parents of A-T patients) have not documented increased spontaneous instability either. For example, in the glycophorin study cited earlier, 23 obligatory A-T heterozygotes did not have significant elevations in their frequency of GPA hemizygous or homozygous variant cells (BIGBEE et al. 1989). More recently, COLE and ARLETT (1994) compared the frequencies of spontaneous *HPRT* mutants among circulating lymphocytes of 16 A-T homozygotes and 19 A-T heterozygotes to those of age-matched controls. The mutation frequency in cells from A-T homozygotes averaged 11-fold higher than controls, but there were no significant differences between heterozygotes and controls.

3.2.1.2 Abnormal Responses to Induced DNA Damage in A-T Heterozygotes

When compared to cells from A-T homozygotes, heterozygote cells have similar, but more subtle, abnormalities in their responses to induced DNA damage. These abnormalities include decreases in survival following exposure to ionizing radiation or radiomimetic drugs, increased sensitivity to damage-induced chromosome and chromatid aberrations, post-irradiation DNA synthesis abnormalities, and prolonged G_2 arrest (CHEN et al. 1978; PATERSON et al. 1979; KIDSON et al. 1982; KINSELLA et al. 1982; KOHN et al. 1982a; SHILOH et al. 1982b; ARLETT and PRIESTLEY 1985; LITTLE and NAGASAWA 1985; PATERSON et al. 1985; NAGASAWA et al. 1985, 1987; ARLETT et al. 1989; ROSIN et al. 1989).

3.2.1.3 Radiosensitivity and the Effects of Dose Rate

The sensitivity of cells from A-T heterozygotes to the killing effects of ionizing radiation and radiomimetic drugs is mild compared to that of A-T homozygotes. Consequently, it sometimes has been difficult to distinguish between A-T heterozygotes and controls when studying the cytotoxic effects of ionizing radiation delivered at typical dose rates. For example, in the first report of radiation sensitivity of A-T heterozygotes, PATERSON et al. (1979) found that the radiation resistance of primary fibroblast lines from three of five A-T heterozygotes was intermediate between control cells and A-T homozygotes' cells. KINSELLA et al. (1982) were unable to confirm these findings when they studied six A-T heterozygotes, but NAGASAWA et al. (1985) found intermediate radiosensitivity in fibroblasts from six different A-T heterozygotes. Similar variability has been reported for the radiosensitivity of peripheral lymphocytes from A-T heterozygotes (e.g., COLE et al. 1988).

Attempts to determine the sensitivity of A-T heterozygotes to radiation-induced chromosomal aberrations have also yielded mixed results. Using high dose-rate irradiation, several groups have reported normal levels of radiation-induced chromosome aberrations in both lymphocytes and fibroblasts (BENDER et al. 1985; NAGASAWA et al. 1985; JONES et al. 1995), while there have been two reports of increased sensitivity to chromatid aberrations induced in G_2 in A-T heterozygote fibroblasts (PARSHAD et al. 1985; SHILOH et al. 1986).

Decreasing the rate of irradiation for control cells improves their survival but does not appreciably affect the survival of cells from A-T homozygotes (PATERSON et al. 1984). Hence, low dose-rates of ionizing irradiation magnify differences between the radioresistance of control cells and A-T homozygotes' cells. In order to reveal subtle defects in radiation response, several groups have studied the survival of A-T heterozygotes cells after low dose-rate irradiation, reasoning that low dose-rate irradiation might accentuate differences in survival between control cells and A-T heterozygotes' cells. This indeed appears to be the case, in that multiple reports using low dose-rate irradiation all have demonstrated statistically significant differences in survival between lymphocytes and fibroblasts from groups of A-T heterozygotes and controls (PATERSON et al. 1985; WEEKS et al. 1991; WAGHRAY et al. 1992; HANNAN et al. 1991, 1994; WEST et al. 1995). For example, WEEKS et al. (1991) used low dose-rate irradiation to study clonal survival of primary fibroblasts from controls, A-T heterozygotes and A-T homozygotes. They found that D_{10} values for ten strains of control fibroblasts averaged 797 cGy, while D_{10} values for nine A-T heterozygotes averaged 496 cGy, and D_{10} values for three A-T homozygotes averaged 165 cGy. The differences between controls and A-T heterozygotes was highly significant ($\chi^2 = 11.14$, $P = 0.0009$). Several of these reports also examined the effects of high dose-rate irradiation on survival. For example, PATERSON et al. (1985) found that there was little difference between the survival of control and A-T heterozygote skin fibroblasts when given a single acute dose of γ-rays (1.1 Gy/min) but that the survival of controls and A-T heterozygotes was significantly different if exposed to chronic low dose-rate irradiation (0.005 Gy/min).

Decreasing the rate of irradiation also protects control cells from radiation-induced chromosomal aberrations, thus magnifying the sensitivity of A-T heterozygotes' cells to radiation-induced chromosomal damage (WAGHRAY et al. 1992; JONES et al. 1995). SCOTT'S group (SCOTT et al. 1993; JONES et al. 1995) and others (WAGHRAY et al. 1992) have taken advantage of this effect to demonstrate that, when compared to controls, A-T heterozygotes have statistically significant elevations in their frequencies of radiation-induced chromosomal aberrations. For example, in a recent study by JONES et al. (1995), yields of dicentric chromosomes in lymphoblasts following 3 Gy γ-rays delivered at a high dose-rate (1.7 Gy/min) averaged 61.6 for 32 control individuals, 75.2 for eight A-T homozygotes, and 60.4 for 19 A-T heterozygotes. When low dose-rate irradiation was used (0.0031 Gy/min), the average yields of dicentric chromosomes dropped to 63 for the A-T homozygotes, 41.5 for the A-T heterozygotes, and 36.0 for the controls. Differences in the responses of heterozygotes and controls was statistically significant only for low dose-rate irradiation ($p < 0.05$).

3.2.1.4 DNA Repair Anomalies in A-T Heterozygotes

Further molecular evidence that *ATM* mutations are not completely recessive comes from studies of radiation-induced double-strand breaks. BLOCHER et al. (1991) used pulsed field gel electrophoresis to examine the number of residual double-strand breaks remaining in genomic DNA following exposure of fibroblasts to low-dose X-rays and found that, on average, cells from five A-T heterozygotes had 50% more residual breaks than controls. This was lower than the threefold elevation seen in A-T homozygotes but suggests that carriers of *ATM* mutations have subtle defects in their repair of certain double-strand breaks. This observation is further supported by work from PANDITA and HITTELMAN (1994) which showed that, using high-dose irradiation and premature chromosome condensation, four A-T heterozygote lymphoblastoid lines sustained increased levels of chromosome damage when compared to controls, even though they had normal survival as well as induction and repair of DNA double-strand breaks following high-dose irradiation.

3.2.1.5 Cell Cycle Checkpoints in A-T Heterozygotes

A cardinal feature of the cells of A-T homozygotes is their lack of cell cycle checkpoints. Although it may lie at the heart of genetic instability in A-T, this aspect of the A-T phenotype has not been thoroughly investigated in A-T heterozygotes. There have been no published studies testing the G_1/S checkpoint in actively growing cells from A-T heterozygotes. However, fibroblasts from A-T heterozygotes undergo an abnormally prolonged delay in entry into S phase following irradiation in G_0 (RUDOLPH et al. 1989).

Several reports have examined the S phase checkpoint in A-T heterozygotes. SHILOH et al. (1982b) found that the average degree of inhibition of DNA synthesis following the exposure of fibroblasts from three A-T heterozygotes to the radiomimetic mutagen neocarzinostatin were smaller that normal but did not find the differences statistically significant. In a more recent study by HANNAN et al. (1994), peripheral lymphocytes and primary fibroblasts from three A-T heterozygotes all had statistically significant defects in the S phase checkpoint, as measured by the presence of "radioresistant" DNA synthesis following exposure to 4 Gy X-rays. However, YOUNG and PAINTER (1989) found that fibroblasts from two obligate A-T heterozygotes had a normal S phase checkpoint following irradiation, and WIENCKE et al. (1992) did not find radioresistant DNA synthesis in lymphocytes from A-T heterozygotes. This suggests that, like other facets of the phenotype, not all A-T heterozygotes may express a defect in their G_1/S checkpoints.

The initial G_2/M checkpoint may be intact in A-T heterozygotes since both primary fibroblasts and peripheral lymphocytes from A-T heterozygotes temporarily pause in G_2 following irradiation (YOUNG and PAINTER 1989; SCOTT et al. 1994). However, several studies have indicated that fibroblasts and lymphoblasts from most, but not all, A-T heterozygotes have a tendency towards delayed G_2 arrest following radiation that is similar, but less marked, than that seen in A-T homozygotes (Bates and Lavin 1989; Rudolph et al. 1989; Lavin et al. 1992, 1994).

3.2.1.6 Dominant Negative *ATM* Alleles as an Explanation for the A-T Heterozygote Phenotype

A molecular explanation of the phenotypic abnormalities of most A-T heterozygotes could be that many *ATM* mutations are not completely recessive, but rather function as weak, dominant negative mutations. ATM proteins may need to form complexes with themselves or other proteins in order to properly function in the damage surveillance signal transduction network. Mutant ATM proteins that are missing key domains (e.g., the PI-3 kinase region) still may participate in complex formation but yet fail to function, thereby inactivating the complex and perturbing the cellular response to DNA damage. The result is a dominant negative phenotype. Those remaining A-T heterozygotes who express no phenotypic abnormalities may harbor null mutations of *ATM*, which would function as completely recessive alleles and have no effect in the heterozygote state.

3.2.2 FA Heterozygotes

As a group, FA heterozygotes express some of the same phenotypic abnormalities as FA homozygotes but in an attenuated form. In vivo abnormalities in obligate FA heterozygotes include statistically significant increases in the frequency of minor skeletal malformations as well as mildly elevated fetal hemoglobin levels (Petridou and Barrett 1990). In addition, blood relatives of FA homozygotes appear to have an excess number of major genitourinary and hand malformations (Gmyrek et al. 1965; Altay et al. 1975; Weshimer and Swift 1982). As discussed above, some FA heterozygotes also may express increased cancer risk.

3.2.2.1 Genetic Instability in FA Heterozygotes

A few studies have found statistically significant differences between the spontaneous frequencies of chromosome aberrations in FA heterozygote cells and controls (Cervenka and Hirsch 1983; Porfirio et al. 1983; Korkina et al. 1992). However, most studies found no significant differences between the two groups (e.g., Cohen et al. 1982; Marx et al. 1983; Rosendorff and Bernstein 1988). FA heterozygotes do not manifest increased in vivo spontaneous mutation frequencies at either the *HPRT* or the glycophorin locus, (Sala-Trepat et al. 1993) but Korkina et al. (1992) found elevated amounts of 8-hydroxy-2′-deoxyguanosine in the genomic DNA of FA heterozygotes, suggesting that they sustain more ROS damage than normal individuals. Lymphocytes from FA heterozygotes express a prolonged G_2 phase when grown in culture, although their cell cycle defect is not as pronounced as that of FA homozygotes (Dutrillaux et al. 1982; Marx and Smith 1989). In addition, caffeine increases the frequency of spontaneous chromosome aberrations in FA heterozygotes to a greater extent than controls (Pincheira et al.

1988), suggesting that a subtle spontaneous genetic instability in FA heterozygotes can be uncovered by loss of G_2 damage-sensitive checkpoint.

In contrast to spontaneous genetic instability, multiple studies have found that, as a group, cells from FA heterozygotes are significantly more sensitive than controls to the clastogenic effects of MMC, HN_2, 8-MOP, and DEB (AUERBACH and WOLMAN 1978; BERGER et al. 1980; AUERBACH et al. 1981; ALIMENA et al. 1983; MARX et al. 1983; PORFIRIO et al. 1983; DALLAPICCOLA et al. 1985; ROSENDORFF and BERNSTEIN 1988). For example, AUERBACH and WOLMAN (1978) found that, following exposure to 10 ng/ml DEB, chromosome breakage in lymphocytes from FA heterozygotes was fourfold greater than that of controls but much less than that induced in cells from FA homozygotes. However, in every study except one (MARX et al. 1983), some controls and FA heterozygotes had responses that overlapped each other, with some reports not even finding group differences (e.g., COHEN et al. 1982; CERVENKA and HIRSCH 1983; ROSENDORFF and BERNSTEIN 1988).

The overlap in behaviour between FA heterozygotes and controls may be the result of genetic heterogeneity. The one study that did not find any overlap between FA heterozygotes and controls in the frequency of DEB-induced chromosome aberrations was a study of 54 heterozygotes and 42 controls from an inbred South African Afrikaner population, where cases of FA most likely are caused by a single FA mutation from a common ancestor (MARX and SMITH 1989). Other studies have found a correlation between the severity of the phenotype within families. For example, WUNDER and FLEISCHER-REISCHMANN (1983) reported that all heterozygotes from three FA families were sensitive to the cytotoxic effects of 8-MOP, while all heterozygotes from three other FA families had normal cellular survival following 8-MOP exposure. DEGAN et al. (1995) found a marked variability in the amount of 8-hydroxy-2′-deoxyguanosine present in the DNA of FA parents but found that this variability was strongly correlated with the amount of 8-hydroxy-2′-deoxyguanosine in the DNA of their affected homozygote children.

3.2.3 BS Heterozygotes

Studies of cells from obligate BS heterozygotes have found normal or near-normal genetic stability, suggesting that *BLM* mutations are completely recessive at the cellular level. These analyses have, in general, focused on chromosome aberrations. Cytogenetic studies have found no differences between BS heterozygotes and controls in the frequency of spontaneous SCEs, quadriradials, or other chromosome aberrations (CHAGANTI et al. 1974; BARTRAM et al. 1976; GERMAN et al. 1977b; KUHN and THERMAN 1979; ROSIN and GERMAN 1985; HOJO et al. 1995). In studies of MN variants at the glycophorin locus in erythrocytes that detected greater than 50-fold elevations in variant frequencies in BS homozygotes, the frequency of variants in three of four BS heterozygotes tested was normal and the fourth near normal (KYOIZUMI et al. 1989; LANGLOIS et al. 1989). An in vitro study of micronuclei formation in epithelial cells from BS heterozygotes found normal frequencies of micronuclei (ROSIN and GERMAN 1985), while a study of micronuclei formation in BS heterozygote fibroblasts grown in culture found threefold increases in spontaneous micronuclei frequencies (FRORATH et al. 1984). The same study did not find increased susceptibility to the induction of micronuclei by mutagens, however.

3.3 What Proportion of All Cancers Occur in Heterozygote Carriers of Chromosome Instability Gene Mutations?

In a 1981 review of the etiology of cancer, DOLL and PETO wrote: "At present, the relevance of genetic susceptibility to the common types of cancer remains obscure." Since then, the inherited components of cancer risk have been much more clearly defined as more than 20 genes responsible for specific familial cancer syndromes have been isolated (reviewed in SANKARANARAYANAN and CHAKRABORTY 1995). The impact of germ line mutations in these genes on the overall incidence of cancer is variable. For example, germ line mutations in *p53* convey a very high risk of developing certain cancers, but the population frequency of *p53* heterozygotes is so low (~0.002%), that only a small portion of human cancers are the result of germ line mutations in this tumor suppressor gene (FORD and EASTON 1995). In contrast, germ line mutations in the *BRCA1* and *BRCA2* genes together account for most cases of familial breast cancer and 4%–8% of all breast cancers, affecting ~500 000 women in the United States alone (ROWELL et al. 1994).

As with other cancer susceptibility genes, the contribution of germ line mutations in the chromosome instability syndrome genes to the overall incidence of cancer is dependent both on the penetrance of heterozygous mutations and on mutant allele frequency. In the following section, estimates for the heterozygote carrier frequency for A-T, FA, and BS mutations will be reviewed, and the potential impact of these germ line mutations on the overall incidence of cancer discussed.

3.3.1 Prevalence of Heterozygote Carriers in the General Population

3.3.1.1 Ataxia Telangiectasia

Taken as a whole, published clinical and laboratory studies reviewed in Section 3.1.1 support the conclusion that A-T heterozygotes are a subpopulation of individuals at increased risk to develop breast and, perhaps, other cancers. Given this conclusion, the frequency of A-T heterozygotes in the general population becomes an important public health consideration. The recent isolation of the *ATM* gene makes direct measurement of carrier frequency possible. However, until such studies are completed, we have to rely on estimates of the frequency of A-T heterozygotes in the general population based on the incidence of diagnosed A-T homozygotes and pedigree analysis of A-T families.

The first published estimate of the prevalence of A-T homozygotes was one in 40 000 live births, based on a survey of elementary school children in Los Angeles (BODER and SEDGWICK 1963). SWIFT et al. (1986), in a retrospective epidemiological study, calculated a lower limit for the mutant gene frequency in the United States Caucasian population of 0.17%, based on the incidence of ascertained cases. Analysis of the incidence of affected cousins in several large pedigrees from the same data set gave a most likely gene frequency of 0.7%, with 95% confidence limits of 0.12%–2%. In a survey of British children born between 1969 and 1976 with A-T, PIPPARD et al. (1988) calculated an incidence of one in 100 000 at birth. In a later study, WOODS et al. (1990) ascertained the prevalence of A-T in the British West Midlands, estimating a birth frequency of one in 300 000. CHESSA et al. (1994), used Dahlberg's formula, together with data from the Italian national registry for A-T patients and Vatican records for consanguineous marriages, to calculate the mutant *ATM* gene frequency among Italians. Their calculated mutant gene frequency, $q = 1.2\% \pm 0.65$, is the highest published estimate and implies an incidence at birth of one in ~7000, suggesting that *ATM* mutations may be unusually prevalent in the Italian population.

As shown in Table 3, one can use gene frequency estimates to estimate the proportion of individuals in the general population who are heterozygote carriers of *ATM* mutations. However, it should be noted that ascertainment of A-T homozygotes in retrospective surveys is incomplete, and gene frequencies based on the prevalence of ascertained cases are almost certainly underestimates. Given the practical difficulties in ascertainment, it is reasonable to assume that a significant proportion of A-T homozygotes were not identified during these surverys. However, published estimates that are based on pedigree analysis appear high. For example, pedigree analysis of the SWIFT et al. data implies that 90% of A-T homozygotes were not ascertained in this survey, an unlikely event. The true

Table 3. Published estimates of mutant *ATM* gene frequency

	BODER and SEDGWICK (1963)	SWIFT et al. (1986)	SWIFT et al. (1986)	PIPPARD et al. (1988)	WOODS et al. (1990)	CHESSA et al. (1994)
Number of A-T patients ascertained	?	231	231	61	11	73
Population surveyed	Los Angeles (US)	United States (Caucasian)	United States (Caucasian)	England	West Midlands (England)	Italy
Basis of estimate	Prevalence in survey	Prevalence in survey	Pedigree analysis	Prevalence in survey	Prevalence in survey	Pedigree analysis
Mutant *ATM* gene frequency (%)	0.5	0.17	0.7	0.32	0.18	1.21
Frequency of mutant *ATM* homozygotes (A-T patients)	~1/40 000	~1/346 000	~1/20 000	~1/97 000	~1/310 000	~1/7 000
Frequency of mutant *ATM* heterozygotes[a] (A-T carriers)	~1/100	~1/300	~1/70	~1/156	~1/277	~1/42

[a]Calculated assuming Hardy-Weinberg equilibrium for a single gene disorder (i.e., no genetic heterogeneity).

gene frequency probably lies between the high and low estimates derived from retrospective population surveys and pedigree analyses.

Previous estimates of the frequency of heterozygote carriers generally have assumed the existence of multiple A-T genes (genetic heterogeneity), based on the finding of at least four A-T complementation groups, as defined by heterodikaryon fusion analyses (JASPERS et al. 1988) and the assumption that each complementation group represented a separate gene. Accordingly, published carrier frequency estimates are higher than they would have been if calculated assuming a single gene disorder.

Although multiple complementation groups represent independent genes for FA and xeroderma pigmentosa, the A-T complementation groups A/B, C, D, and E recently were found to be caused by mutations in the *ATM* gene (SAVITSKY et al. 1995). In light of this finding that a single gene is responsible for essentially all classic cases of A-T, the published estimates of heterozygote carrier frequency need to be revised downwards, as has been done in Table 3. These revised carrier frequencies range from 0.34% to 2.42%, with a reasonable average of about 1%.

These estimates, together with the data summarized above that suggest that carriers of *ATM* mutations are at an approximately fourfold increased risk for breast cancer, lead to the conclusion that roughly 3%–5% of women with breast cancer are heterozygotes for mutations in the *ATM* gene. This estimate is similar to that of EASTON (1994a) and suggests that, in the United States, ~7000 female heterozygote carriers of *ATM* mutations develop breast cancer each year. As to the possible contribution of heterozygote *ATM* carriers to other types of cancer, the epidemiological and laboratory data reviewed in Sections 3.1.1 and 3.2.1 suggest a slight predisposition to gastric carcinomas and other malignancies, but the epidemiological data are not consistent (see Section 3.1.1), and any estimate based on these published studies would have a great deal of uncertainty.

3.3.1.2 Fanconi Anemia

Fanconi anemia has been reported in all major ethnic groups (ALTER 1994a) but is not as common as A-T, with the combined incidence of FA homozygotes from all complementation groups estimated to be ~1/350 000 (SWIFT 1971; SCHROEDER et al. 1976; SWIFT et al. 1980; SASAKI 1989). Given the existence of at least three and perhaps as many as five different FA genes, the combined heterozygote carrier frequency for all FA genes may be as high as one in 130. Specific ethnic groups may be at higher risk. For example, a recent survey of 3100 Jewish individuals for *IVS4* and *322delG* mutations in the *FACC* gene found a carrier frequency for the *IVS4* mutation of 1/89 in Ashkenazi Jews (VERLANDER et al. 1995), and ROSENDORFF et al. (1987) has estimated a 1/77 carrier frequency for the South African Afrikaners FA mutation, presumably due to a founder effect in this isolated population.

The studies summarized in Section 3.2.2 indicate that many FA heterozygotes express, to a lesser degree, the phenotypic abnormalities seen in FA homozygotes, including an increased sensitivity to the induction of chromosomal aberrations by bifunctional DNA cross-linking agents. These data suggest that at least some FA mutations are not completely recessive and predict that many FA heterozygotes may have a mild degree of genetic instability in vivo. A consequence of this in vivo genetic instability might be an increased incidence of cancer. However, given the results of the few epidemiological surveys of FA families that have been conducted (see Sect. 3.1.2), this increase is likely to be minimal for most FA mutations. As a result, despite the fact that the carrier frequency of all FA mutations approaches that of *ATM* mutations in the general population, it is likely that the contribution of FA heterozygote carriers to the overall incidence of cancer is low. The possibility remains that a few FA mutations may carry a significant risk of cancer for heterozygote carriers, e.g., the FA family reported by HILL (1976). However, the contributions of such mutations to the overall cancer burden are likely to be minor.

3.3.1.3 Bloom Syndrome

Bloom Syndrome occurs in many different ethnic groups (GERMAN and PASSARGE 1989) but is considerably rarer than A-T and FA, with the United States incidence of BS homozygotes estimated to be 1/6 330 000 (GERMAN and TAKEBE 1989). A study of BS in Israel documented increased occurrence in Ashkenazi Jews and estimated the homozygote gene freqency in that population as 1/60 000, with a carrier frequency of ~1/125 (GERMAN et al. 1977a).

The studies reviewed in Section 3.2.3 indicate that cells from BS heterozygotes have normal or near-normal genetic stability, suggesting that *BLM* mutations are completely recessive at the cellular level. This is consistent with anecdotal reports that BS heterozygotes are not at a high risk to develop cancer (GERMAN 1995). Given the lack of evidence for increased genetic instability and cancer in BS heterozygotes and their relatively low carrier frequency, the impact of germ line mutations in the *BLM* gene on the overall incidence of inherited cancers would appear to be minimal or non-existent.

3.3.2 Do Germ Line Mutations in the Chromosome Instability Genes Play a Role in Familial Cancers?

Cancer is a common enough condition that occurrence of the same cancer in more than one first-degree relative frequently occurs by chance. However, it has long been recognized that familiar clustering of cancer can be the result of germ line mutations in tumor suppressor genes. For example, p53 is responsible for many cases of the Li-Fraumeni cancer predisposition syndrome, and hereditary non-polyposis colon cancers (HNPCC) are due to loss of function mutations in the *hMLH1*, *hMSH2*, *hpMS1*, and *hpMS2* genes that control the repair of DNA mismatches (FISHEL et al. 1993; LEACH et al. 1993; BIRCH et al. 1994; BRONNER et al. 1994; NICOLAIDES et al. 1994; PAPADOPOULOS et al. 1994). Given that these genes which control genetic stability have been implicated in familial cancer, one might expect that the mutations in one or more chromosome instability syndrome genes also would be associated with autosomal dominant familial cancer predisposition syndromes. However, there is little to support this possibility and, in the case of

ATM, several studies suggest that the gene does not play a major role in familial breast cancer.

3.3.2.1 *ATM* Mutations and Familial Breast Cancer

In a recent study, Cortessis et al. (1993) examined 64 families in which two or more individuals had had premenopausal bilateral breast cancer. They tested for possible linkage of the 11q23 marker *DRD2* to dominant transmission of breast cancer predisposition in these families. When all families were included in the analysis, there was strong evidence against linkage (i.e., a LOD score of –6.08 at $\Theta = 0.00001$). Even if families with linkage to 17q21 or a family history of ovarian cancer were excluded, there still was no evidence for linkage to DRD2. However, given that *DRD2* is ~7 cM from the *ATM* gene, this study could not, by itself, definitively exclude linkage to the *ATM* gene itself.

Wooster et al. (1993) typed five 11q markers in individuals from 16 families in which there were at least two affected sisters with cancer before the age of 45, together with at least one other relative with breast cancer. Families with individuals affected with ovarian cancer were excluded so as to eliminate families in which *BRCA1* mutations may be responsible for cancer predisposition. Multipoint linkage analysis of these families yielded a calculated LOD score of –1.34 for the linkage between the *ATM* locus and breast predisposition in these families. The authors concluded that *ATM* mutations do not make a major contribution to familial breast cancer.

Since surveys of A-T families consistently show an increased relative risk of developing breast cancer in *ATM* heterozygotes, one might expect that mutations in the *ATM* gene would play a role in familial breast cancer. Yet the studies of breast cancer families summarized above did not demonstrate linkage to the *ATM* locus on 11q23. As Easton (1994a) has pointed out, both studies actually support the same conclusion: *ATM* mutations do not play a major role in familial breast cancer.

The apparent contradiction between studies of familial breast cancer families and A-T families can be resolved if one considers the penetrance of heterozygous *ATM* mutations. Using the published data for the occurrence of breast cancer amongst blood relatives of A-T homozygotes, Easton (1994b) has calculated the risk of developing breast cancer for A-T heterozygotes to be 11% by age 50 and 30% by age 70. As summarized in Table 4, this is a relatively low penetrance compared to the *BRCA1, BRCA2,* or *p53* genes. One effect of this low penetrance is that first-degree relatives of A-T heterozygotes should have only a modest increase in their relative risk of developing breast cancer. As a result, *ATM* mutations should account for relatively few "breast cancer families," which, by the usual definition, are families in which clusters of multiple individual cancers have occurred.

Easton et al. (1993) have estimated that, given their low penetrance, *ATM* mutations should be responsible for only 3.5% of familial breast cancers, while *BRCA1* and *BRCA2* mutations together account for ~90%. However, because

Table 4. Risk of developing breast cancer in heterozygote carriers of selected genes

Germ line mutation	Heterozygote carrier frequency (%)	Risk of breast cancer by age 50 years (%)	70 years (%)	Proportion of all breast cancer patients < 50 years (%)	> 50 years (%)	Proportion of familial breast cancer patients (%)	Sources
No mutation		1.4	5				(CLAUS et al. 1991)
ATM	~1	11	30	7	2	3.5	(EASTON 1994b)
BRCA1	~0.2–0.3	50	85	5	1	~50	(EASTON et al. 1993)
BRCA2	~0.5?		63			40	(SANKARANARAYANAN and CHAKRABORTY 1995)
p53	~0.002	40	90	1	< 1		(FORD and EASTON 1995)
HRAS	~4–6	3	8				(KRONTARIS et al. 1993)

ATM mutations are more frequent in the general population than mutations in these highly penetrant genes, germ line *ATM* mutations may give rise to many "sporadic" tumors. The net result is that *ATM* mutations may be responsible for at least as many cases of breast cancer in the general population as germ line mutations in *BRCA1*, *BRCA2*, or *p53*.

4 Do Somatic Mutations in the Chromosome Instability Genes Play a Role in Sporadic Tumor Development?

Cancer is thought to be the end result of a multistep process that includes the acquisition of mutations in multiple genes that control cellular behaviour (NOWELL 1976; VOGELSTEIN and KINZLER 1993). As such, cancer can be considered to be a somatic genetic disease in which a population of cells undergoes clonal evolution as the result of sustaining multiple genetic changes that activate oncogenes and inactivate tumor suppressor genes. For certain childhood tumors (e.g., retinoblastoma), as few as two mutations may be sufficient to give rise to a tumor (reviewed in Goodrich and Lee 1990). However, for common adult cancers, three to six mutations may be necessary to acquire the phenotypic characteristics of a fully malignant tumor (LOEB 1991; VOGELSTEIN and KINZLER 1993).

Somatic human cells normally maintain tight control of genomic stability, with spontaneous mutation rates calculated to be $1.2–1.4 \times 10^{-10}$ mutations/base pair per cell generation (summarized in LOEB 1991). This has led to the hypothesis that loss of normal control of genomic stability is a necessary step in carcinogenesis that allows clonal populations of cells to accumulate a sufficient number of mutations to acquire a fully malignant phenotype (NOWELL 1976; LOEB 1991). This hypothesis is supported by observations that genetic instability is expressed by most tumors (LOEB 1991; HARTWELL 1992; LANE 1992; VOGELSTEIN and KINZLER 1993) and that genes involved in maintaining genetic stability frequently are found to be inactivated in tumor cells (e.g., *p53*, *hMLH1*, *hMSH2*, *hpMS1*, and *hpMS2*).

Given the loss of control of genomic stability in chromosome instability syndrome patients, one might expect that the chromosome instability syndrome genes would, like *p53* and the DNA mismatch repair genes, play key roles as guardians of the genome. If so, then loss of function of the chromosome instability syndrome genes may be a frequent event in the development of sporadic common cancers. With the cloning of the *ATM*, *FACC*, and *BLM* genes, it should be possible to provide definitive answers to this question in the near future. However, published studies of loss of heterozygosity of genetic markers in sporadic tumors already provide indirect evidence that at least some of the chromosome instability syndrome genes function as tumor suppressor genes that are targets for inactivation during the development and progression of common sporadic cancers.

4.1 Loss of Heterozygosity Studies

Loss of heterozygosity (LOH) studies use RFLPs, microsatellites, and other genetic polymorphisms to map regional deletions and amplifications of chromosomes in tumors. Regional alterations in the genome are frequent events in tumorigenesis, presumably because they can unmask recessive mutations in tumor suppressor loci (Cavenee et al. 1983). Specific tumors tend to have consistent patterns of genomic alteration. For example, regions with frequent LOH in breast cancer include 1p, 3p, 11p15, 11q, 16q, 17p, 17q, and 18q (reviewed in Gudmundsson et al. 1995). The consistency of these genomic alterations in tumors has facilitated the mapping and subsequent cloning of several tumor suppressor genes that play major roles in the development of common cancers. These studies also can be used to indict candidate genes in the development of specific malignancies as a prelude to further study of their potential as tumor suppressor genes.

4.1.1 11q23 LOH and the *ATM* Gene

The best case that chromosome instability syndrome genes are targets for inactivation during the development of common cancers comes from multiple studies that have found that loss of heterozygosity at 11q22-23 loci is a frequent event in sporadic breast, ovarian, colon, and lung cancers as well as malignant melanoma.

4.1.1.1 Breast Cancer

Several studies have documented that loss of heterozygosity at 11q22-23 is a frequent finding in sporadic breast cancers. For example 40% of a combined total of 182 sporadic tumors examined in four independent studies had lost heterozygosity at *D11S35*, a microsatellite locus ~8 cM centromeric to the *ATM* locus (Carter et al. 1994; Hampton et al. 1994a; Gudmundsson et al. 1995; Winqvist et al. 1995).

Although most LOH studies provide only a very rough molecular estimate as to the location of the relevant locus that is lost during tumor development, several groups have attempted to specify an interstitial location for the 11q23 locus that is commonly deleted in sporadic breast carcinomas. In an analysis of 62 sporadic breast cancers, Carter et al. (1994) found that 39% of the tumors had lost 11q23 markers, with the common overlapping region of LOH having *D11S35* and *D11S29* as the centromeric and telomeric limits, respectively. This narrowed the 11q23 region of LOH to an ~30 cM region that includes the A-T locus. Based on an independent analysis of 44 breast cancers. HAMPTON et al. (1994a) defined the telomeric limit for deletion as proximal to *APOC3*, 3 cM centromeric to *D11S29* (UHRHAMMER et al. 1994). In a more recent study, TOMLINSON et al. (1995) found breast cancers that had LOH at *NCAM* but not *DRD2*. This moves the telomeric limit for the common region of LOH to 6 cM distal of the *ATM* locus. Finally, using polymorphic markers tightly linked to the *ATM* locus, NEGRINI et al. (1995) found that 50% (14/28) of tumors examined had lost heterozygosity at *D11S1818*, a marker that lies less than 1 cM telomeric to *ATM* (UHRHAMMER et al. 1994). NEGRINI et al. (1995) concluded that the region of minimal 11q23 deletion in these tumors was an ~9 cM area that included *ATM* and excluded another 11q23 candidate tumor suppressor gene, *ALL-1*.

Inactivation of the *ATM* locus may be an early event in the development of many breast carcinomas in that KORETH et al. (1995) found that 23% of in situ breast carcinomas had 11q23 LOH. Loss of *ATM* function may be an indicator of a poor long-term prognosis, since 11q23 LOH has been associated both with metastatic disease (KORETH et al. 1995) and with decreased survival after metastasis (WINQVIST et al. 1995). Interestingly, CARTER et al. (1994) found a correlation between LOH at 17p13 and 11q23 in the tumors they analyzed, suggesting that loss of both *ATM* and *p53* frequently occurs in the same tumor (although compare GUDMUNDSSON et al. 1995). Loss of both of these tumor suppressor genes could explain the poor prognosis seen in breast cancers with 11q23 LOH in that the resultant tumor would be both genetically unstable from a lack of multiple cell cycle checkpoints and resistant to treatment due to the loss of p53-dependent DNA damage-induced apoptosis.

4.1.1.2 Other Tumors

Although evidence for regional 11q23 chromosomal loss involving the *ATM* locus is strongest for breast cancers, regional deletions involving 11q23 also have been documented for several other solid tumors. For example two cytogenetic studies of colorectal cancers found frequent deletions of 11q in these tumors (MULERIS et al. 1990; KONSTANTINOVA et al. 1991). A combined cytogenetic and molecular study of 39 colorectal carcinomas by KELDYSH et al. (1993) found that 59% of the tumors had loss of 11q material, with the smallest region of overlap being 11q22-23. GUSTAFSON et al. (1994) studied colorectal carcinomas for LOH at the *DRD2* locus, a marker approximately 8 cM telomeric to *ATM* (UHRHAMMER et al. 1995). They found that 34% of 68 tumors had undergone LOH at *DRD2*, a proportion that was higher than their background for random markers (8%-23%) but lower than that seen for the 68% LOH seen for the *p53* region of 17p. Taken together, these studies suggest that loss of *ATM* function may occur in many, but not the majority, of colon carcinomas.

Malignant melanoma also may be a cancer in which LOH at 11q23 is a common event. Tomlinson et al. (1993) examined 52 malignant melanomas for LOH at four candidate regions, including 11q23, 17p13.3, 17p13.1, and 17q22. A high frequency of LOH occurred only at the 11q23 marker (*D11S29*), where 66% of the tumors showed LOH. In a subsequent study of 24 melanoma patients, Herbst et al. (1995) used microsatellite repeats to define a critical region of 11q23 LOH for melanoma that includes the *ATM* locus. However, given that the 11q23 critical region defined by Herbst et al. is 51 cM wide, it is far from certain that the *ATM* locus is a target for genetic loss in melanoma.

Cervical carcinomas have a high degree of karyotypic instability, but the instability is non-random, with many breakpoints clustered around 11q23-q25 (Sreekantaiah et al. 1991). Hampton et al. (1994b) studied 32 cervical carcinomas and found that 52% had LOH of the 11q23 marker *D11S144*. Bethwaite et al. (1995) also examined cervical carcinomas for 11q LOH and found that 16 out of 53 tumors had lost heterozygosity at the *D11S29* locus, which lies telomeric to *ATM*, while preserving heterozygosity at 11p13 loci.

To date, there have been two reports of 11q23 LOH in ovarian and lung cancers. Iizuka et al. (1995) analyzed 79 lung cancers and found that 11q24 LOH was a frequent event. This finding was confirmed by a recent study from Croce's group which examined 76 adenocarcinomas of the lung and found LOH of 11q in 63%, with three distinctive regions affected by loss, including a 4-Mb critical region of minimal loss that encompasses the *ATM* gene (Rasio et al. 1995). Ovarian tumors also have been examined for 11q23 LOH (Weitzel et al. 1994; Pejovic 1995), but the evidence of an association between 11q23 LOH and the development of ovarian cancer is relatively weak. For example, Weitzel et al. (1994) examined LOH at multiple loci in a series of 27 primary ovarian cancers and found high frequencies of loss (>50%) at markers near the *APC* gene on 5q21 and *BRCA1* locus on 17q21 but only 29% LOH at the 11q23 marker *D11S533*.

These LOH studies provide broad, but indirect, support for the idea that loss of *ATM* function plays a role in the development of many common tumors. With the recent isolation of the *ATM* gene, confirmation of these LOH studies should be forthcoming in the near future, and a more accurate picture of the role of *ATM* gene inactivation in tumorigenesis will emerge.

4.1.2 LOH and FA Genes

There is little evidence to suggest that LOH involving the *FACC* gene is a frequent event during tumor development. Bladder carcinomas, large cell lymphomas, and esophageal carcinomas all have been shown to undergo LOH for 9q, but recent fine mapping demonstrates that the critical regions of loss for these tumors center around 9q34, distal to the *FACC* locus (Mori et al. 1994; Chaganti et. al. 1995;

HABUCHI et al. 1995). LOH involving the region of 9q containing the *FACC* gene, 9q22.3, is a frequent event in basal cell carcinomas and medulloblastomas (ALBRECHT et al. 1994; SHANLEY et al. 1995). However, interpretation of LOH studies involving 9q22.3 is complicated by the fact that three potential cancer predisposition genes are within several megabases of each other at 9q22.3: the *FACC* gene, the gene for xeroderma pigmentosum group A (*XPAC*), and the Gorlin syndrome gene (MORRIS and REIS 1994). Basal cell carcinomas and medulloblastomas are characteristic tumors of Gorlin syndrome (BALE et al. 1994), and LOH at 9q22 is frequently found in those basal cell carcinomas and medulloblastomas that occur in Gorlin syndrome patients (ALBRECHT et al. 1994; SCHOFIELD et al. 1995; SHANLEY et al. 1995; LEVANAT et al. 1996). Fine mapping has demonstrated that, for at least some basal cell carcinomas, the area of loss does not include the *FACC* gene (SHANLEY et al. 1995). Taken together, these observations suggest that the gene at 9q22.3 that is a target for inactivation in basal cell carcinomas and medulo-blastomas is the Gorlin syndrome gene, not *FACC*.

Loss of heterozygosity for the region of 16q that contains the *FA(A)* gene, 16q23-q24, is a frequent event in sporadic breast carcinomas (DENG et al. 1994; HARADA et al. 1994; ITO et al. 1995; SKIRNISDOTTIR et al. 1995). In addition, 16q24 LOH has been reported to be frequent in hepatocellular carcinomas (YUMOTO et al. 1995). The critical region for LOH at 16q23-24 has not been as narrowly defined as the critical region for LOH at 11q23, but these studies are consistent with the possibility that the *FA(A)* gene is a specific target for inactivation during the development of breast and liver carcinomas and suggest that further investigation is warranted into the possibility that inactivation of the *FA(A)* gene occurs during the development of these tumors.

Many sporadic solid tumors have been shown to exhibit frequent LOH for loci on 3p, including breast carcinomas (CHEN et al. 1994), cervical carcinomas (KOHNO et al. 1993; KARLSEN et al. 1994; MULLOKANDOV et al. 1996), esophageal carcinomas (MORI et al. 1994; OGASAWARA et al. 1995), non-small cell lung carcinomas (reviewed in THIBERVILLE et al. 1995), mesotheliomas (LU et al. 1994), and head and neck squamous carcinomas (EL-NAGGAR et al. 1993; PARTRIDGE et al. 1994). The critical regions for these tumors range from 3p13 to 3p26, suggesting that there are multiple tumor suppressor genes on the short arm of chromosome 3. The microcell-mediated chromosome transfer experiments that resulted in the initial mapping of the *FA(D)* locus to 3p did not permit assignment of a specific map location for *FA(D)* (WHITNEY et al. 1995). Consequently, it is not yet possible to critically evaluate which, if any, of the sporadic tumors that demonstrate 3p LOH might inactivate the *FA(D)* gene during the course of tumor development.

4.1.3 15q LOH and the *BLM* Gene

Although one study found that nine out of 30 (35%) sporadic breast carcinomas demonstrated loss of the 15q (DEVILEE et al. 1991), the 15q26 region that contains the *BLM* gene does not appear to be a part of the genome that undergoes frequent LOH in sporadic tumors. Molecular analysis of the *BLM* gene in tumor tissue may yet provide evidence for the inactivation of *BLM* during tumorigenesis. However, at this time, LOH studies suggest that *BLM* is not a frequent target for inactivation during tumor development.

5 Cancer Screening and Treatment

5.1 Lessons for Cancer Screening

As discussed above, A-T and FA heterozygotes may, in aggregate, represent several percent of the general population. Because many of them may have abnormal responses to DNA damage, A-T and FA heterozygotes may represent a group of individuals whose risks for developing cancer differs from that of the general population. Much needs to be learned before cancer screening issues in A-T and FA carriers can be fully resolved. However, our knowledge of A-T has progressed to the point where several points regarding cancer screening in A-T heterozygotes can begin to be addressed.

5.1.1 Screening for A-T Heterozygotes

Since A-T heterozygotes are at an increased risk to develop cancer and may have adverse reactions to cancer therapy, it would be useful to have a screening test to identify this high-risk subpopulation. The published studies reviewed above support the conclusion that groups of A-T heterozygotes, *on the average*, express more spontaneous genetic instability and greater sensitivity to the toxic effects of ionizing radiation than groups of control individuals. Multiple attempts have been made to develop screening tests for A-T heterozygotes. However, reliable screening tests for identifying A-T heterozygotes based on functional abnormalities have yet to be developed due to consistent overlap between the most radioresistant A-T heterozygotes and the most radiosensitive "normal" individuals (e.g., Weeks et al. 1991).

This overlap may be due to a combination of two factors. First, A-T heterozygotes with apparently normal radioresistance may carry null mutations or other *ATM* alleles that do not interfere with the function of the remaining normal *ATM* gene but, instead, act like true recessive mutations. Second, given the population frequency of A-T heterozygotes (Section 3.3.1.1), it is likely that some radiosensitive normal individuals actually may be heterozygote carriers of *ATM* mutations, while others may carry mutations in other genes that affect radiation resistance.

Cloning the *ATM* gene has opened up the possibility of developing molecular screening tests for A-T heterozygotes. However, initial analyses of *ATM* mutations in A-T homozygotes suggest that there are many "private" mutations scattered throughout the gene (Randal 1995b; Savitsky et al. 1995; R.A. Gatti 1996, personal communication). Given the large size of the *ATM* cDNA and the existence of multiple different mutations, it may be difficult, using current molecular technology, to develop a molecular screening test for A-T heterozygotes that will detect enough *ATM* mutations to be clinically useful. However, since many of the known mutations appear to code for protein truncations, it may be possible to develop screening tests that detect truncated ATM protein fragments. Protein truncation screening would not identify true null alleles. However, if dominant-negative alleles play a role in the cancer risk of A-T heterozygotes, then truncation screening may be of practical use since it could identify those *ATM* alleles that code for abnormal proteins that function as dominant-negative alleles.

5.1.2 Mammography and Radiation-Induced Breast Cancer in A-T Heterozygotes

Annual mammograms generally are recommended for women under age < 50 who are at high risk to develop breast cancer (KERLIKOWSKE et al. 1993; VOGEL 1994). A-T heterozygotes would appear to be a high-risk group that could benefit from such screening. However, in SWIFT et al.'s (1991) prospective analysis of cancer in blood relatives of A-T homozygotes, there was a significant correlation between breast cancer and a history of one are more fluoroscopic examinations of the chest, back, or abdomen ($p < 0.005$), with a relative risk of 5.8 for those individuals with a history of radiation exposure. This led SWIFT et al. (1991) to suggest that physicians consider alternatives to mammography in A-T heterozygotes. It should be noted, however, that SWIFT currently is neither for nor against mammographic screening for A-T heterozygotes (RANDAL 1995a).

The recommendation against mammography provoked much discussion and some criticism of the study's methodology by members of the cancer genetics, epidemiology, and radiology communities (e.g., BOICE and MILLER 1992; GRAY 1992; KULLER AND MODAN 1992; RANDAL 1995b). In particular, several individuals pointed out an apparent inconsistency between the relative risk of cancer seen in the irradiated group and their incremental exposure to radiation (BOICE and MILLER 1992; LAND 1992). Given that the women with breast cancer in the study already had 40–50 mGy cumulative lifetime exposures from background radiation, the incremental exposures from diagnostic radiation (1–9 mGy) appeared to be much too low to account for the 5.8-fold increase in relative risk seen in the irradiated group of A-T blood relatives. Although the study population was the largest group of A-T families yet analyzed for cancer risk, there still were only small numbers of blood relatives and spouse controls that developed cancer. To date, there have been no further published epidemiological studies that have studied links between radiation exposure and cancer risk in A-T heterozygotes. However, new light may be shed on this controversy now that investigators are using *ATM* gene analysis to reexamine this question (RANDAL 1995b).

Another way to approach this issue is to consider what conclusions can be drawn from published estimates of the risks of induced cancer from mammography in the general population and from the studies of the radiation responses of A-T heterozygotes reviewed above. Typical radiation exposures to the breasts from mammography are in the range, on average, of 2–3 mGy (MILLER et al. 1989; BOICE and MILLER 1992), compared to the 1–2 mGy per year exposure from background radiation sources (National Research Council 1990). Based on long-term studies of Hiroshima and Nagasaki survivors, as well as smaller studies of women exposed to diagnostic and therapeutic radiation, it has been estimated that, for the general population, the risk of breast cancer from radiation doses of 10 mGy is ~ 1 in 10 000 per year (National Research Council 1990). Norman and Withers (1993) have used these figures to calculate that the additional lifetime risk for breast cancer resulting from annual mammography is about 1.5% for the general population. The sensitivity of cells from A-T heterozygotes to radiation-induced genetic alterations averages about 25%–60% higher than that of controls (e.g., Jones et al. 1995). As pointed out by Norman and Withers (1993), this suggests that the lifetime risk of breast cancer in A-T heterozygotes from mammography is less than twofold higher than controls, perhaps 2%–3%. This compares to an estimated lifetime risk of developing breast cancer in A-T heterozygotes of > 30% (Easton 1994b). Given that annual mammography reduces the mortality of breast cancer through early detection by ~ 30% or more (Rennert 1991), the benefits of annual mammography would appear to outweigh the risk of mammography-induced cancer in A-T heterozygotes by age 40, if not earlier.

5.2 Lessons for Cancer Treatment

A recurring problem in oncology is how to devise an effective treatment regime that can be safely tolerated by the cancer patient. A-T and FA genes, like other genes that control genetic stability, may be frequent targets for inactivation during tumorigenesis. Loss of function of these genes decreases the genetic stability of the malignant cells but at the same time affects their sensitivity to radiation and/or chemotherapeutic agents. For example, inactivation of *ATM* could play a role in the sensitivity of many tumor cells to induction of apoptosis by ionizing radiation, radiomimetic chemicals, and topoisomerase inhibitors (Dive et al. 1992).

Although much remains to be learned, it eventually may be possible to develop individualized treatment strategies that are tailored to exploit the phenotypic differences between a patient's malignant cells and their normal somatic cells that are caused by loss of function of chromosome instability syndrome genes or other tumor suppressor genes. For example, sporadic tumors that have lost *ATM* function may, like cells from A-T homozygotes, be quite sensitive to the induction of apoptosis by ionizing radiation and radiomimetic drugs, while those tumors that have inactivated their FA genes may become sensitive to killing by MMC and other cross-linking agents. New drugs also could be explicitly screened for their ability to exploit the differences between malignant cells and normal tissues in DNA repair, cell cycle checkpoint control, and cytokine regulation (Marshall 1996).

5.2.1 Chromosome Instability Homozygotes Are at Risk for Adverse Treatment Reactions

Although their overall numbers are small, A-T, FA, and BS homozygotes make up a subpopulation of cancer patients with relatively bleak prognoses due in part to their poor responses to therapeutic radiation and chemotherapy (see Sections 2.1, 2.4, and 2.7). Care must be taken to modify standard regimes so as to minimize iatrogenic damage, but reduced therapeutic regimes can be tolerated by these individuals (e.g., EYRE et al. 1988; PASSARGE 1991; ALTER 1994a).

5.2.2 Chromosome Instability Syndrome Heterozygotes as Subpopulation at Risk for Adverse Reactions to Treatment

Given the abnormal responses of A-T and FA heterozygotes to DNA-damaging agents, they could, in theory, represent a significant fraction of cancer patients who have adverse responses to radiation and chemotherapy. Experimental evidence is currently lacking for this possibility in FA heterozygotes; however, studies of radiosensitive cancer patients and A-T heterozygotes provide indirect support for this hypothesis.

5.2.2.1 A-T Heterozygotes and Radiosensitive Cancer Patients

In theory, sufficiently high doses of ionizing radiation could eliminate any tumor. However, in practice, the therapeutic radiologist must limit radiation treatment so as not to induce unacceptable damage to surrounding normal tissues. Unfortunately, apparently normal individuals vary widely in their radiosensitivity (LITTLE et al. 1988; GENTNER and MORRISON 1989; TURESSON 1990). As a result, radiation treatments are typically mild enough so as to prevent adverse reactions in the occasional radiosensitive individual. A consequence of this approach is that many patients are given less radiation than they could reasonably tolerate, thus diminishing the probability that their radiation therapy will be curative.

It has been proposed that the radiosensitive "tail" of the population distribution of radiation responses may actually represent a subgroup of individuals who are radiosensitive as the result of mutations in specific genes (NORMAN et al. 1988). If such individuals could be identified and given modified therapy, then the efficiency of radiation therapy for the majority of cancer patients could be significantly improved by giving them more intensive radiation without the fear of an unacceptable number of adverse reactions (NORMAN et al. 1988; WEST et al. 1991)

For these reasons, multiple attempts have been made to correctly identify radiosensitive cancer patients using in vitro assays that measure the radiosensitivity of fibroblasts, lymphocytes, or lymphoblastoid cells. The results have been mixed. Most studies have demonstrated a group correlation between in vitro radiosensitivity and adverse reactions to radiotherapy (LOEFFLER et al. 1990; BURNET et al. 1992; LAVIN et al. 1994; JONES et al. 1995; RAMSAY and BIRRELL 1995). However, this has not always been the case for studies using survival following high-dose irradiation as their measure of radiosensitivity (e.g., Begg et al. 1993). As with studies of A-T heterozygotes, those in vitro studies that either have focused on cell cycle abnormalities or used low dose-rate irradiation have shown the most promise in identifying which cancer patients have in vivo radiosensitivities.

For example, Lavin et al. (1994) found that 20% of 104 unselected breast cancer patients had lymphocytes with radiation-induced G_2 delays in the A-T heterozygote range, compared to 8% of lymphoblasts from 24 age-matched controls ($p < 0.001$). In this study, an abnormally high fraction of cells in G_2 correlated with more severe clinical reactions to radiation, as well as a family history of breast cancer and more aggressive tumors. More recently, West et al. (1995) determined the survival, after low dose-rate irradiation, of lymphocytes from 13 breast cancer patients who had had severe early or late tissue reactions to radiation therapy. The average D_0 for this group of over-reactors was similar to that for 13 A-T heterozygotes. Both groups were significantly more sensitive than a group of 25 controls without cancer as well as eight breast cancer patients who had had no adverse radiation reactions ($p < 0.01$). The same investigators also were able to find statistically significant differences between over-reactors and controls using low dose-rate irradiation and radiation-induced chromosome aberrations (Jones et al. 1995). Unfortunately, as with the case of A-T heterozygote identification, a small degree of overlap between the in vitro radiosensitivity of over-reactors and controls precludes the clinical application of these functional assays at this time.

Taken as a whole, these reports indicate that intrinsic radiosensitivity may cause some individuals to suffer adverse reactions to therapeutic radiation. In these studies, cells from cancer patients with clinical radiosensitivity had in vitro responses to radiation that were similar to those of cells from A-T heterozygotes. This has led several investigators to suggest that some radiosensitive cancer patients may be unusually sensitive to radiation therapy because they are A-T heterozygotes (Lavin et al. 1994; Jones et al. 1995). However, Swift (1994) has suggested, based on anecdotal evidence, that A-T heterozygotes do not have marked in vivo sensitivity to therapeutic radiation. This issue may have to be resolved by molecular analysis of the *ATM* gene in selected radiosensitive cancer patients.

6 Conclusion

The three diseases discussed above demonstrate the critical role that genetic instability plays in the development of cancer and illustrate the wide range of nuclear and cytoplasmic processes that can affect the stability of the genome. As discussed in this review, a great deal of progress has been made in the last few years towards understanding the molecular pathology of these chromosome instability syndromes, with recent efforts focusing on cloning the genes responsible for these syndromes. Now, with the identification of several of the genes, we are entering a new era of intense study that will focus on the pathobiology of these syndromes as well as the role played by the chromosome instability syndrome genes in DNA metabolism, cell cycle checkpoints, genetic instability, and programmed cell death. These studies should test the predictions of our current models for the chromosome instability syndromes, further our understanding of basic biological processes, shed light on the origin and development of cancer, and improve our diagnosis and treatment of this formidable disease.

References

Adler-Brecher B, Zhang Q, Flit Y, Gray DL, Beaver HA, Reid JE, Auerbach AD (1992) Prenatal diagnosis of Fanconi anemia in monozygotic twin boys with discordant phenotype. Am J Hum Genet 51: A251

Agamanolis DP, Greenstein JI (1979) Ataxia-telangiectasia: report of a case with Lewy bodies and vascular abnormalities within cerebral tissue. J Neuropathol Exp Neurol 39: 475–489

Aguilar MJ, Kamoshita S, Landing BH, Boder E, Sedgwick RP (1968) Pathological observations in ataxia-telangiectasia. A report on 5 cases. J Neuropathol Exp Neurol 27: 659–676

Albrecht S, von Deimling A, Pietsch T, Giangaspero F, Brandner S, Kleihues P, Wiestler OD (1994) Microsatellite analysis of loss of heterozygosity on chromosome 9q, 11p and 17p in medulloblastomas. Neuropathol Appl Neurobiol 20: 74–81

Alimena G, Avvisati G, De Cuia MR, Gallo E, Novelli G, Dallapiccola B (1983) Retrospective diagnosis of a Fanconi's anemia patient by dyepoxybutane (DEB) test results in parents. Haematologica 68: 97–103

Allen C, Meyn MS (1995) A low threshold for mutagen-induced apoptosis in Fanconi anemia cells. Am J Hum Genet 57: A46

Altay C, Sevgi Y, Pirnar T (1975) Letter: Fanconi's anemia in offspring of patient with congenital radial and carpal hypoplasia. N Engl J Med 293: 151 152

Alter BP (1992) Fanconi's anemia. Current concepts. Am J Pediatr Hematol Oncol 14: 170–176

Alter BP (1994a) Clinical features of Fanconi's anemia. In: Young NS, Alter BP (eds) Aplastic anemia: acquired and inherited. Saunders, Philadelphia, pp 275–309

Alter BP (1994b) Inherited bone marrow failure syndromes: introduction. In: Young NS, Alter BP (eds) Aplastic anemia: acquired and inherited. Saunders, Philadelphia, pp 271–274

Alter BP (1994c) Pathophysiology of Fanconi's anemia. In: Young NS, Alter BP (eds) Aplastic anemia: acquired and inherited. Saunders, Philadelphia, 310–324

Alter BP, Scalise A, McCombs J, Najfeld V (1993) Clonal chromosomal abnormalities in Fanconi's anemia: what do they really mean? Br J Haematol 85: 627–630

Amromin GD, Boder E, Teplitz R (1979) Ataxia-telangiectasia with a 32 year survival. A clinicopathological report. J Neuropathol 38: 621–643

Anderson CW (1993) DNA damage and the DNA-activated protein kinase. Trends Biochem Sci 18: 433–437

Arlett CF, Harcourt SA (1978) Cell killing and mutagenesis in repair-defective human cells. In: Hanawalt PC, Friedberg EC, Fox CF (eds) DNA repair mechanisms. Academic, New York, pp 633–636

Arlett CF, Harcourt SA (1980) Survey of radiosensitivity in a variety of human cell strains. Cancer Res 40: 926–932

Arlett CF, Priestley A (1985) An assessment of the radiosensitivity of ataxia-telangiectasia heterozygotes. In: Gatti RA, Swift M (eds) Ataxia-telangiectasia: genetics, neuropathology, and immunology of a degenerative disease of childhood. Liss, New York, pp 101–109

Arlett CF, Green MH, Priestley A, Harcourt SA, Mayne LV (1988) Comparative human cellular radiosensitivity. I. The effect of SV40 transformation and immortalisation on the gamma-irradiation survival of skin derived fibroblasts from normal individuals and form ataxia-telangiectasia patients and heterozygotes. Int J Radiat Biol 54: 911–928

Arlett CF, Cole J, Green MHL (1989) Radiosensitive individuals in the population. In: Baverstock KF, Stather JW (eds) Low dose radiation: biological bases of risk assessment. Taylor and Francis, London, pp 240–252

Auerbach AD (1993) Fanconi anemia diagnosis and the diepoxybutane (DEB) test (editorial). Exp Hematol 21: 731–733
Auerbach AD (1995) Fanconi anemia. Dermatol Clin 13: 41–49
Auerbach AD, Allen RG (1991) Leukemia and preleukemia in Fanconi anemia patients. A review of the literature and report of the International Fanconi Anemia Registry. Cancer Genet Cytogenet 51: 1–12
Auerbach AD, Wolman SR (1976) Susceptibility of Fanconi's anemia fibroblasts to chromosome damage by carcinogens. Nature 261: 494–496
Auerbach AD, Wolman SR (1978) Carcinogen–induced chromosome breakage in Fanconi's anaemia heterozygous cells. Nature 271: 69–71
Auerbach AD, Wolman S (1979) Carcinogen-induced chromosome breakage in chromosome instability syndromes. Cancer Genet Cytogenet 1: 21–28
Auerbach AD, Adler B, Chaganti RS (1981) Prenatal and postnatal diagnosis and carrier detection of Fanconi anemia by a cytogenetic method. Pediatrics 67: 128–135
Auerbach AD, Rogatko A, Schroeder-Kurth TM (1989) International Fanconi Anemia Registry: relation of clinical symptoms to diepoxybutane sensitivity. Blood 73: 391–396
Aurias A, Dutrillaux B (1986) Acquired inversions in human leucocytes. Ann Genet (Paris) 29: 203–206
Aurias A, Dutrillaux B, Buriot D, Lejeune J (1980) High frequencies of inversions and translocations of chromosome 7 and 14 in ataxia telangiectasia Mutat Res 69: 369–374
Aurias A, Antoine J-L, Assathiany R, Odievre M, Dutrillaux B (1985a) Radiation sensitivity of Bloom's syndrome lymphocytes during S and G_2 phases. Cancer Genet Cytogenet 16: 131–136
Aurias A, Dutrillaux AM, Dutrillaux B, Herpin F, Lamoliatte E, Lombard M, Muleris M, Paravatou M, Prieur M, Prod'homme M, Sportes M, Viegas-Pequignot E, Voloboriev V (1985b) Inversion (14) (q12qter) or (q11.2q32.3): the most frequently acquired rearrangement in lymphocytes. Hum Genet 71: 19–21
Austin CA, Sng JH, Patel S, Fisher LM (1993) Novel HeLa topoisomerase II is the II beta isoform: complete coding sequence and homology with other type II topoisomerases. Biochim Biophys Acta 1172: 283–291
Averbeck D, Averbeck S (1985) Genotoxic effects of mono-and bifunctional furocoumarins in yeast: involvement of DNA photoaddition and oxygen dependent reactions. In: Bensasson RV, Jori G, Land FD, Truscott TG (eds) Primary photo-processes in biology and medicine. Plenum, New York, pp 295–300
Bagby GC, Jr., Segal GM, Auerbach AD, Onega T, Keeble W, Heinrich MC (1993) Constitutive and induced expression of hematopoietic growth factor genes by fibroblasts from children with Fanconi anemia. Exp Hematol 21: 1419–1426
Bagnara GP, Bonsi L, Strippoli P, Ramenghi U, Timeus F, Bonifazi F, Bonafe M, Tonelli R, Bubola G, Brizzi MF et al (1993) Production of interleukin 6, leukemia inhibitory factor and granulocyte-macrophage colony stimulating factor by peripheral blood mononuclear cells in Fanconi's anemia. Stem Cells (Dayt) 11 [Suppl 2]: 137–143
Baker SJ, Reddy EP (1996) Transducers of life and death: TNF receptor superfamily and associated proteins. Oncogene 12: 1–10
Bale AE, Gailani MR, Leffell DJ (1994) Nevoid basal cell carcinoma syndrome. J Invest Dermatol 103: 126S–130S
Bargman GJ, Shahidi NT, Gilbert EF, Opitz JM (1977) Studies of malformation syndromes of man. XLVII. Disappearance of spermatogonia in the Fanconi anemia syndrome. Eur J Pediatr 125: 163–168
Bartram CR, Koske-Westphal T, Passarge E (1976) Chromatid exchanges in ataxia telangiectasia, Bloom syndrome, Werner syndrome, and xeroderma pigmentosum. Ann Hum Genet 40: 79–86
Bates PR, Lavin MF (1989) Comparison of gamma-radiation-induced accumulation of ataxia telangiectasia and control cells in G_2 phase. Mutat Res 218: 165–170
Battista JR, Donnelly CE, Ohta T, Walker GC (1990) The SOS response and induced mutagenesis. Prog Clin Biol Res 340A: 169–178
Baumann T (1951) Konstitutionelle Panmyelopathose mit multiplen Abartungen (Fanconi-Syndrome) Ann Paediatr (Basel) 177: 142–174
Beamish H, Lavin MF (1994) Radiosensitivity in ataxia-telangiectasia: anomalies in radiation-induced cell cycle delay. Int J Radiat Biol 65: 175–184
Begg AC, Russell NS, Knakan H, Lebesque JV (1993) Lack of correlation of human fibroblast radiosensitivity in vitro with early skin reactions in patients undergoing radiotherapy. Int J Radiat Biol 64: 393–405
Bender BA, Rary JM, Kale RP (1985) G_2 chromosomal radiosensitivity in ataxia telangiectasia lymphocytes. Mutat Res 152: 39–47

Bennett CB, Rainbow AJ (1988) Delayed expression of enhanced reactivation and decreased mutagenesis of UV-irradiated adenovirus in UV-irradiated ataxia telangiectasia fibroblasts. Mutagenesis 3: 389–395
Berger R, Coniat ML (1989) Cytogenetic studies in Fanconi anemia induced chromosomal breakage and cytogenetics. In: Schroeder-Kurth TM, Auerbach AD, Obe G (eds) Fanconi anemia. Springer, Berlin Heidelberg New York, pp 93–99
Berger R, Bernheim A, Le Coniat M, Vecchione D, Schaison G (1980) Effect of chlormethin chlorhydrate on the chromosomes in Fanconi's anemia: application to diagnosis and detection heterozygotes. CR Seances Acad Sci D 290: 457–459
Bertazzoni U, Scovassi AI, Stefanini M, Giulotto E, Spadari S, Pedrini AM (1978) DNA polymerases alpha beta and gamma in inherited diseases affecting DNA repair. Nucleic Acids Res 5: 2189–2196
Bethwaite PB, Koreth J, Herrington CS, McGee JO (1995) Loss of heterozygosity occurs at the D11S29 locus on chromosome 11q23 in invasive cervical carcinoma. Br J Cancer 71: 814–818
Bigbee WL, Langlois RG, Swift M, Jensen RH (1989) Evidence for an elevated frequency of in vivo somatic cell mutations in ataxia telangiectasia. Am J Hum genet 44: 402–408
Bigbee WL, Jensen RH, Grant SG, Langlois RG, Olsen DA, Auerbach A (1991) Evidence for elevated in vivo somatic mutation at the glycophorin A locus in Fanconi anemia. Am J Hum Genet 49 [Suppl]: 446
Bigelow SB, Rary JM, Bender MA (1979) G_2 chromosomal radiosensitivity in Fanconi's anemia. Mutat Res 63: 189–199
Birch JM, Hartley AL, Tricker KJ, Prosser J, Condie A, Kelsey AM, Harris M, Morris PH, Binchy A, Crowther D, Craft AW, Enden OB, Evans DGR, Thompson E, Mann JR, Martin J, Mitchell ELD, Santibáñez-Koref MF (1994) Prevalence and diversity of constitutional mutations in the p53 gene among 21 Li-Fraumeni families. Cancer Res 54: 1298–1304
Blocher D, Sigut D, Hannan MA (1991) Fibroblasts from ataxia telangiectasia (AT) and AT heterozygotes show an enhanced level of residual DNA double-strand breaks after low dose-rate gamma-irradiation as assayed by pulsed field gel electrophoresis. Int J Radiat Biol 60: 791–802
Bloom D (1954) Congenital telangiectatic erythema resembling lupus erythematosus in dwarfs. Am J Dis Child 88: 754–758
Blunt T, Finnie NJ, Taccioli GE, Smith GC, Demengeot J, Gottlieb TM, Mizuta R, Varghese AJ, Alt FW, Jeggo PA et al (1995) Defective DNA-dependent protein kinase activity is linked to V(D)J recombination and DNA repair defects associated with the murine scid mutation. Cell 80: 813–823
Boder E., Sedgwick RP (1957) Ataxia-telangiectasia. A familial syndrome of progressive cerebellar ataxia, oculocutaneous telangiectasia and frequent pulmonary infection. A preliminary report on 7 children, an autopsy, and a case history. Univ South Calif Med Bull 9: 15–28
Boder E, Sedgwick RP (1958) Ataxia-telangiectasia. A familial syndrome of progressive cerebellar ataxia, oculocutaneous telangiectasia and frequent pulmonary infection. Pediatrics 21: 526–554
Boder E, Sedgwick RP (1963) Ataxia-telangiectasia. A review of 101 cases In: Walsh G (ed) Little club clinics in developmental medicine. Heinemann, London, pp 110–118
Boice JD, Miller RW (1992) Risk of breast cancer in ataxia-telangiectasia (letter). N Engl J Med 326: 1357–1358
Borresen AL, Andersen TI, Tretli S, Heiberg A, Moller P (1990) Breast cancer and other cancers in Norwegian families with ataxia-telangiectasia. Genes Chromosom Cancer 2: 339–340
Borsellino N, Belldegrun A, Bonavida B (1995) Endogenous interleukin 6 is a resistance factor for cis-diamminedichloroplatinum and etoposide-mediated cytotoxicity of human prostate carcinoma cell lines. Cancer Res 55: 4633–4639
Boulikas T (1995) Phosphorylation of transcription factors and control of the cell cycle. Crit Rev Eukaryotic Gene Expr 5: 1–77
Brainard E, Herzing LBK, Bainton J, Meyn MS (1991) A common feature of the chromosome instability syndromes: increased spontaneous intrachromosomal mitotic recombination. Am J Hum Genet 49A: 449
Bronner CE, Baker SM, Morrison PT, Warren G, Smith LG, Lescoe MK, et al (1994) Mutations in the DNA mismatch repair gene homolog hMLH1 is associated with hereditary nonpolyposis colon cancer. Nature 368: 258–261
Bubley GJ, Schnipper LE (1987) Effects of Bloom's syndrome fibroblasts on genetic recombination and mutagenesis of herpes simplex virus type 1. Somat Cell Mol Genet 13: 111–117
Burger RM, Drlica D, Birdsall B (1994) The DNA cleavage of iron bleomycin. J Biol Chem 269: 25978–25985
Burnet NG, Nyman J, Turesson I, Wurm R, Yarnold JR, Peacock JH (1992) Prediction of normal-tissue tolerance to radiotherapy from in vitro cellular radiation sensitivity. Lancet 339: 1590–1591

Butturini A, Gale RP, Verlander PC, Adler-Brecher B, Gillio AP, Auerbach AD (1994) Hematologic abnormalities in Fanconi anemia: an International Fanconi Anemia Registry study. Blood 84: 1650–1655

Buul PPV, Rooij DGD, Zandman IM, Grigorova M, Duyn-Goedhart AV (1995) X-ray-induced chromosomal aberrations and cell killing in somatic and germ cells of the scid mouse. Int J Radiat Biol 67: 549–555

Canman CE, Wolff AC, Chen CY, Fornace AJ Jr, Kastan MB (1994) The p53-dependent G_1 cell cycle checkpoint pathway and ataxia-telangiectasia. Cancer Res 54: 5054–5058

Carter SL, Negrini M, Baffa R, Gillum DR, Rosenberg AL, Schwartz GF, Croce CM (1994) Loss of heterozygosity at 11q22–q23 in breast cancer. Cancer Res 54: 6270–6274

Cavenee WK, Dryja TP, Phillips RA, Benedict WF, Godbout R, Gallie B, Murphree A, Strong LC, White RL (1983) Expression of recessive alleles by chromosomal mechanisms in retinoblastoma. Nature 305: 779–784

Cervenka J, Hirsch BA (1983) Cytogenetic differentiation of Fanconi anemia, idiopathic aplastic anemia, and Fanconi anemia heterozygotes. Am J Med Genet 15: 211–223

Chaganti RSK, Schonberg S, German J (1974) A manyfold increase in sister chromatid exchanges in Bloom's syndrome lymphocytes. Proc Natl Acad Sci USA 71: 4508–4512

Chaganti SR, Gaidano G, Louie DC, Dalla-Favera R, Chaganti RS (1995) Diffuse large cell lymphomas exhibit frequent deletions in 9p21–22 and 9q31–34 regions. Genes Chromosom Cancer 12: 32–36

Chan JY, Becker FF, German J, Ray JH (1987) Altered DNA ligase I activity in Bloom's syndrome cells. Nature 325: 357–359

Chen LC, Matsumura K, Deng G, Kurisu W, Ljung BM, Lerman MI, Waldman FM, Smith HS (1994) Deletion of two separate regions on chromosome 3p in breast cancers. Cancer Res 54: 3021–3024

Chen M, Cumming R, Krasnoshtein F, Savoia A, Lightfoot J, Santos C, Parker L, Wong J, Joyner A, Buchwald M (1995) Molecular genetics of Fanconi anemia. J Cell Biochem 21A: 272

Chen PC, Lavin MF, Kidson C, Moss D (1978) Identification of ataxia telangiectasia heterozygotes, a cancer prone population. Nature 274: 484–486

Chessa L, Lisa A, Fiorani O, Zei G (1994) Ataxia-telangiectasia in Italy: genetic analysis. Int J Radiat Biol 66: S31–33

Chu J-Y, Ho JE, Monteleone PL, O'Conner DM (1979) Technicium colloid bone marrow imaging in Fanconi's anemia. Pediatrics 64: 952–954

Claus EB, Risch N, Thompson WD (1991) Genetic analysis of breast cancer in the cancer and steroid hormone study. Am J Hum Genet 48: 232–242

Cohen MM, Levy HP (1969) Chromosome instability syndromes. Adv Hum Genet 18: 43–149

Cohen MM, Shaham M, Dagan J, Shmueli E, Kohn G (1975) Cytogenetic investigations in families with ataxia-telangiectasia. Cytogenet Cell Genet 15: 338–356

Cohen MM, Simpson SJ, Honig GR, Maurer HS, Nicklas JW, Martin AO (1982) The identification of Fanconi anemia genotypes by clastogenic stress. Am J Hum Genet 34: 794–810

Cole J, Arlett CF (1994) Cloning efficiency and spontaneous mutant frequency in circulating T-lymphocytes in ataxia-telangiectasia patients. Int J Radiat Biol 66: S123–131

Cole J, Arlett CF, Green MH, Harcourt SA, Priestley A, Henderson L, Cole H, James SE, Richmond F (1988) Comparative human cellular radiosensitivity. II. The survival following gamma-irradiation of unstimulated (G_0) T-lymphocytes, T-lymphocyte lines, lymphoblastoid cell lines and fibroblasts from normal donors, from ataxia-telangiectasia patients and from ataxia-telangiectasia heterozygotes. Int J Radiat Biol 54: 929–943

Cornforth MN, Bedford JS (1985) On the nature of a defect in cells from individuals with ataxia-telangiectasia. Science 227: 1589–1591

Cortessis V, Ingles S, Millikan R, Diep A, Gatti RA, Richardson L, Thompson WD, Paganini-Hill A, Sparkes RS, Haile RW (1993) Linkage analysis of DRD2, a marker linked to the ataxia-telangiectasia gene, in 64 families with premenopausal bilateral breast cancer. Cancer Res 53: 5083–5086

Costa ND, Thacker J (1993) Response of radiation-sensitive human cells to defined DNA breaks. Int J Radiat Biol 64: 523–529

Cowdell RH, Phizackerly PJR, Pyke DA (1995) Constitutional anemia (Fanconi's syndrome) and leukemia in two brothers. Blood 10: 788–801

Cox R, Masson WK, Debenham PG, Webb MBT (1984) The use of recombinant DNA plasmids for the determination of DNA-repair and recombination in cultured mammalian cells. Br J Cancer 46 [Suppl VI] : 67–72

Dallapiccola B, Magnani M, Novelli G, Mandelli F (1984) Increased activity of glutathione S-transferase and fast decay of reduced glutathione in Fanconi's anemia erythrocytes. Acta Haematol 71: 143–144
Dallapiccola B, Porfirio B, Mokini V, Alimena G, Isacchi G, Gandini E (1985) Effects of oxidants and antioxidants on chromosomal breakage in Fanconi anemia lymphocytes. Hum Genet 69: 62–65
Day RS, Giuffrida AS, Dingman CW (1975) Repair by human cells of adenovirus-2 damaged by psoralen plus near ultraviolet light treatment. Mutat Res 33: 311–320
Dean SW, Fox M (1983) Investigation of the cell cycle response of normal and Fanconi's anaemia fibroblasts to nitrogen mustard using flow cytometry. J Cell Sci 64: 265–279
Debenham PG, Webb MB, Stretch A, Thacker J (1988) Examination of vectors with two dominant, selectable genes for DNA repair and mutation studies in mammalian cells. Mutat Res 199: 145–158
Degan P, Bonassi S, De Caterina M, Korkina LG, Pinto L, Scopacasa F, Zatterale A, Calzone R, Pagano G (1995) In vivo accumulation of 8-hydroxy-2′-deoxyguanosine in DNA correlates with release of reactive oxygen species in Fanconi's anaemia families. Carcinogenesis 16: 735–741
Dehazya P, Sirover MA (1986) Regulation of hypoxanthine DNA glycosylase in normal human and Bloom's syndrome fibroblasts. Cancer Res 46: 3756–3761
Deng GR, He LW, Lin BY (1994) Loss of heterozygosity at different chromosomes in patients with breast cancer. Chung Hua I Hsueh Tsa Chih 74: 31–34
Devilee P, van Vliet M, Van Sloun P, Kuipers Dijkshoorn N, Hermans J, Perason PL, Cornelisse CJ (1991) Allelotype of human breast carcinoma: a second major site for loss of heterozygosity is on chromosome 6q. Oncogene 6: 1705–1711
Dive C, Evans CA, Whetton AD (1992) Induction of apoptosis – new targets for cancer chemotherapy. Semin Cancer Biol 3: 417–427
Doll R, Peto R (1981) The causes of cancer: quantitative estimates of avoidable risks of cancer in the United States today. JNCI 66: 1191–1308
Dosik H, Hsu LY, Todaro GJ, Lee SL, Hirschhorn K, Selirio ES, Alter AA (1970) Leukemia in Fanconi's anemia: cytogenetic and tumor virus susceptibility studies. Blood 36: 341–352
Dritschilo A, Brennan T, Weichselbaum RR, Mossman KL (1984) Response of human fibroblasts to low dose rate gamma irradiation. Radiat Res 100: 387–395
Duckworth-Rysiecki G, Taylor AMR (1985) Effects of ionizing radiation of cells from Fanconi's anemia patients. Cancer Res 45: 416–420
Dulic V, Kaufmann WK, Wilson SJ, Tlsty TD, Lees E, Harper JW, Elledge SJ, Reed SI (1994) p53-dependent inhibition of cyclin-dependent kinase activities in human fibroblasts during radiation-induced G_1 arrest. Cell 76: 1013–1023
Dutrillaux B, Aurias A, Dutrillaux AM, Buriot D, Prieur M (1982) The cell cycle of lymphocytes in Fanconi anemia. Hum Genet 62: 327–332
Eady JJ, Peacock JH, McMillan TJ (1992) Host cell reactivation of gamma-irradiated adenovirus 5 in human cell lines of varying radiosensitivity. Br J Cancer 66: 113–118
Easton D, Ford D, Peto J (1993) Inherited susceptibility to breast cancer. Cancer Surv 18: 95–113
Easton DF (1994a) Cancer risks in A-T heterozygotes. Int J Radiat Biol 66: S177–182
Easton DF (1994b) The inherited component of cancer. Br Med Bull 50: 527–535
el-Deiry WS, Harper JW, O'Connor PM, Velculescu VE, Canman CE, Jackman J, Pietenpol JA, Burrell M, Hill DE, Wang Y (1994) WAF1/CIP1 is induced in p53-mediated G_1 arrest and apoptosis. Cancer Res 54: 1169–1174
el-Naggar AK, Lee MS, Wang G, Luna MA, Goepfert H, Batsakis JG (1993) Polymerase chain reaction-based restriction fragment length polymorphism analysis of the short arm of chromosome 3 in primary head and neck squamous carcinoma. Cancer 72: 881–886
Ellis NA, Groden J, Ye T-Z, Straughen J, Lennon DJ, Ciocci S, Proytcheva M, German J (1995a) The Bloom's syndrome gene product is homologous to recQ helicases. Cell 83: 655–666
Ellis NA, Lennon DJ, Proytcheva M, Alhadeff B, Henderson EE, German J (1995b) Somatic intragenic recombination within the mutated locus BLM can correct the high sister-chromatid exchange phenotype of Bloom syndrome cells. Am J Hum Genet 57: 1019–1027
Enoch T, Norbury C (1995) Cellular responses to DNA damage: cell-cycle checkpoints, apoptosis and the roles of p53 and ATM. Trends Biochem Sci 20: 426–430
Evans HJ, Adans AC, Clarkson JM, German J (1978) Chromosome aberrations and unscheduled DNA synthesis in X- and UV-irradiated lymphocytes from a boy with Bloom's syndrome and a man with xeroderma pigmentosum. Cytogenet Cell Genet 20: 124–140
Evans HJ, Ray JH, German J (1983) Bloom's syndrome: Evidence for an increased mutation frequency in vivo. Science 221: 851–853

Evans MK, Bohr VA (1994) Gene-specific DNA repair of UV-induced cyclobutane pyrimidine dimers in some cancer-prone and premature-aging human syndromes. Mutat Res 314: 221–231
Eyre JA, Gardner-Medwin D, Summerfield GP (1988) Leukoencephalopathy after prophylactic radiation for leukaemia in ataxia telangiectasia. Arch Dis Child 63: 1079–1080
Fanconi G (1927) Familiäre infantile perniziösartige Anämie (perniziöses Blutbild und Konstitution). Jahrb Kinderheilkd 117: 257–280
Fanconi G (1967) Familial constitutional panmyelopathy, Fanconi's anemia (F.A.). Semin Hematol 4: 233–240
Feigin RD, Vietti TJ, Wyatt RG, Kaufmann DG, Smith CH (1970) Ataxia telangiectasia with granulocytopenia. J Pediatr 77: 431–438
Fendrick JL, Hallick LM (1984) Psoralen photoinactivation of herpes simplex virus: monoadduct and cross-link repair by xeroderma pigmentosum and Fanconi's anemia cells. J Invest Dermatol 83: 96s–101s
Fishel R, Lescoe MK, Rao MRS, Copeland NG, Jenkins JA, Barber J, Kane M, Kolodner R (1993) The human mutator gene homology MSH2 and its association with hereditary nonpolyposis colon cancer. Cell 75: 1027–1038
Ford D, Easton DF (1995) The genetics of breast and ovarian cancer. Br J Cancer 72: 805–812
Fornace AJ Jr, Little JB, Weichselbaum RR (1979) DNA repair in a Fanconi's anemia fibroblast cell strain. Biochem Biophys Acta 561: 99–109
Fritz E, Herzing L, Elsea S, Meyn MS (1996) A novel human cDNA that complements the ataxia-telangiectasia phenotype is homologous to the S. cerevisiae Top3 topoisomerase. Proc Natl Acad Sci USA (in press)
Frorath B, Schmidt-Preuss U, Siemers U, Zollner M, Rudiger HW (1984) Heterozygous carriers for Bloom syndrome exhibit a spontaneously increased micronucleus formation in cultured fibroblasts. Hum Genet 67: 52–55
Fujiwara Y (1982) Defective repair of mitomycin C crosslinks in Fanconi's anemia and loss in confluent normal human and xeroderma pigmentosum cells. Biochim Biophys Acta 699: 217–225
Fujiwara Y, Tatsumi M (1977) Cross-link repair in human cells and its possible defect in Fanconi's anemia cells. J Mol Biol 113: 635–649
Galloway SM, Evans HJ (1975) Sister chromatid exchange in human chromosomes from normal individuals and patients with ataxia telangiectasia. Cyto Cell Genet 15: 17–29
Gangloff S, McDonald JP, Bendixen C, Arthur L, Rothstein R (1994) The yeast type I topoisomerase Top3 interacts with SGS1, a DNA helicase homolog: a potential eukaryotic reverse gyrase. Mol Cell Biol 14: 8391–8398
Garriga S, Crosby WH (1959) The incidence of leukemia in families of patients with hypoplasia of the marrow. Blood 14: 1008–1014
Gentner NE, Morrison B (1989) Determination of the proportion of persons in the population who exhibit abnormal sensitivity to ionising radiation low dose radiation. Taylor and Francis, London, pp 253–262
German J (1964) Cytological evidence for crossing-over in vitro in human lymphoid cells. Science 144: 298–301
German J (1969) Bloom's syndrome. I. Genetical and clinical observations in the first twenty-seven patients. Am J Hum Genet 21: 196–227
German J (1974) Bloom's syndrome. II. The prototype of human diseases predisposing to chromosome instability and cancer. In: German J (ed) Chromosomes and cancer. Wiley, New York
German J (1983) Patterns of neoplasia associated with the chromosome-breakage syndromes. In: German J (ed) Chromosome mutation and neoplasia. Liss, New York, pp 97–134
German J (1993) Bloom syndrome: a Mendelian prototype of somatic mutational disease. Medicine (Baltimore) 72: 393–406
German J (1995) Bloom's syndrome. Dermatol Clin 13: 7–18
German J, Passarge E (1989) Bloom's syndrome. XII. Report from the Registry for 1987. Clin Genet 35: 57–69
German J, Schonberg S (1980) Bloom's syndrome. IX. Review of cytological and biochemical aspects. In: Gelboin HV, MacMahon B, Matsushima T, Sugimura T, Takayama S, Takebe H (eds) Genetic and environmental factors in experimental and human cancer. Japan Scientific Societies Press, Tokyo, pp 175–186
German J, Takebe H (1989) Bloom's syndrome. XIV. The disorder in Japan. Clin Genet 35: 93–110
German J, Archibald R, Bloom D (1965) Chromosomal breakage in a rare and probably genetically determined syndrome of man. Science 148: 506–507

German J, Bloom D, Passarge E, Fried K, Goodman RM, Katzenellenbogen I, Laron Z, Legum C, Levin S, Wahrman J (1977a) Bloom's syndrome. VI. The disorder in Israel and an estimation of the gene frequency in the Ashkenazim. Am J Hum Genet 29: 553–562

German J, Schonberg S, Louie E, Chaganti RSK (1977b) Bloom's syndrome. IV. Sister chromatid exchanges in lymphocytes. Am J Hum Genet 29: 248–255

German J, Roe AM, Leppert MF, Ellis NA (1994) Bloom syndrome: an analysis of consanguineous families assigns the locus mutated to chromosome band 15q26.1. Proc Natl Acad Sci U S A 91: 6669–6673

Giampietro PF, Adler-Brecher B, Verlander PC, Pavlakis SG, Davis JG, Auerbach AD (1993) The need for more accurate and timely diagnosis in Fanconi anemia: a report from the International Fanconi Anemia Registry. Pediatrics 91: 1116–1120

Giampietro PF, Verlander PC, Maschan A et al (1994) Fanconi anemia: a model for somatic gene mutation during development. Am J Med Genet 52: 36

Gibbons B, Scott D, Hungerford JL, Cheung KL, Harrison C, Attard-Montalto S, Evans M, Birch JM, Kingston JE (1995) Retinoblastoma in association with the chromosome breakage syndromes Fanconi's anaemia and Bloom's syndrome: clinical and cytogenetic findings. Clin Genet 47: 311–317

Gille JJ, Wortelboer HM, Joenje H (1987) Antioxidant status of Fanconi anemia fibroblasts. Hum Genet 77: 28–31

Glanz A, Fraser FC (1982) Spectrum of anomalies in Fanconi anaemia. J Med Genet 19: 412–416

Gluckman E, Devergie A, Dutreix J (1983) Radiosensitivity in Fanconi anaemia: application to the conditioning regimen for bone marrow transplantation. Br J Haematol 54: 431–440

Gluckman E, Auerbach AD, Horowitz MM, Sobocinski KA, Ash RC, Bortin MM, Butturini A, Camitta BM, Champlin RE, Friedrich W et al (1995) Bone marrow transplantation for Fanconi anemia. Blood 86: 2856–2862

Gmyrek D, Otto FM, Sylla-Rapoport I (1965) On the familial occurrence of Fanconi anemia and thrombocytopenia with malformations (remarks on the therapy of Fanconi anemia). Monatsschr Kinderheilkd 113: 542–552

Goodrich DW, Lee WH (1990) The molecular genetics of retinoblastoma. Cancer Surv 9: 529–564

Gotoff SP, Amirmokri E, Liebner EJ (1967) Neoplasia, untoward response to X-irradiation, and tuberous sclerosis. Am J Dis Child 114: 617–625

Gottlieb TM, Jackson SP (1993) The DNA-dependent protein kinase: requirement for DNA ends and association with Ku antigen. Cell 72: 131–141

Gray JE (1992) Increased incidence of cancer in a small subset of the population: a new obstacle to screening mammography? Radiology 185: 285–286

Greenwell PW, Kronmal SL, Porter SE, Gassenhuber J, Obermaier B, Petes TD (1995) TEL1, a gene involved in controlling telomere length in S. cerevisiae, is homologous to the human ataxia telangiectasia gene. Cell 82: 823–829

Groden J, German J (1992) Bloom's syndrome. XVIII. Hypermutability at a tandem-repeat locus. Hum Genet 90: 360–367

Groden J, Nakamura Y, German J (1990) Molecular evidence that homologous recombination occurs in proliferating human somatic cells. Proc Natl Acad Sci USA 87: 4315–4319

Grompe M, Low M, Riefsteck C, Olson S, Whitney MA (1995) Generation and phenotypic characterization of mice disrupted for the Fanconi anemia group C gene. Am J Hum Genet 57: A51

Gudmundsson J, Barkardottir RB, Eiriksdottir G, Baldursson T, Arason A, Egilsson V, Ingvarsson S (1995) Loss of heterozygosity at chromosome 11 in breast cancer: association of prognostic factors with genetic alterations. Br J Cancer 72: 696–701

Gustafson CE, Young J, Leggett B, Searle J, Chenevix-Trench G (1994) Loss of heterozygosity on the long arm of chromosome 11 in colorectal tumours. Br J Cancer 70: 395–397

Habuchi T, Devlin J, Elder PA, Knowles MA (1995) Detailed deletion mapping of chromosome 9q in bladder cancer: evidence for two tumour suppressor loci. Oncogene 11: 1671–1674

Hall EJ, Marchese MJ, Astor MB, Morse T (1986) Response of cells of human origin, normal and malignant, to acute and low dose rate irradiation. Int J Radiat Oncol Biol Phys 12: 655–659

Hall JD, Scherer K (1981) Repair of psoralen-treated DNA by genetic recombination in human cells infected with herpes simplex virus. Cancer Res 41: 5033–5038

Hallahan DE, Beckett MA, Kufe D (1990) The interaction between recombinant human tumor necrosis factor and radiation in 13 human tumor cell lines. Int J Radiat Oncol Biol Phys 19: 69–74

Hampton GM, Mannermaa A, Winquist R, Alavaikko M, Blanco G, Taskinen PJ, Kiviniemi H, Newsham I, Cavenee WK, Evans GA (1994a) Loss of heterozygosity in sporadic human breast carcinoma: a common region between 11q22 and 11q23.3. Cancer Res 54: 4586–4589

Hampton GM, Penny LA, Baergen RN, Larson A, Brewer C, Liao S, Busby-Earle RM, Williams AW, Steel CM, Bird CC et al (1994b) Loss of heterozygosity in cervical carcinoma: subchromosomal localization of a putative tumor-suppressor gene to chromosome 11q22–q24. Proc Natl Acad Sci USA 91: 6953–6957

Hand R, German J (1975) A retarded rate of DNA chain growth in Bloom's syndrome. Proc Natl Acad Sci USA 72: 758–762

Hang B, Yeung AT, Lambert MW (1993) A damage-recognition protein which binds to DNA containing interstrand cross-links is absent or defective in Fanconi anemia, complementation group A, cells. Nucleic Acids Res 21: 4187–4192

Hannan MA, Greer W, Smith BP, Sigut D, Ali MA, Amer MH (1991) Skin fibroblast cell lines derived from non-Hodgkin's-lymphoma (NHL) patients show increased sensitivity to chronic gamma irradiation. Int J Cancer 47: 261–266

Hannan MA, Kunhi M, Einspenner M, Khan BA, al-Sedairy S (1994) Post-irradiation DNA synthesis inhibition and G_2 phase delay in radiosensitive body cells from non-Hodgkin's lymphoma patients: an indication of cell cycle defects. Mutat Res 311: 265–276

Harada Y, Katagiri T, Ito I, Akiyama F, Sakamoto G, Kasumi F, Nakamura Y, Emi M (1994) Genetic studies of 457 breast cancers. Clinicopathologic parameters compared with genetic alterations. Cancer 74: 2281–2286

Hari KL, Santerre A, Sekelsky JJ, McKim KS, Boyd JB, Hawley RS (1995) The mei-41 gene of D. melanogaster is a structural and functional homolog of the human ataxia telangiectasia gene. Cell 82: 815–821

Hartley KO, Gell D, Smith GC, Zhang H, Divecha N, Connelly MA, Admon A, Lees-Miller SP, Anderson CW, Jackson SP (1995) DNA-dependent protein kinase catalytic subunit: a relative of phosphatidylinositol 3-kinase and the ataxia telangiectasia gene product. Cell 82: 849–856

Hartwell LH (1992) Defects in a cell cycle checkpoint may be responsible for the genomic instability of cancer cells. Cell 71: 543–546

Hartewell LH, Kastan MB (1994) Cell cycle control and cancer. Science 266: 1821–1828

Harvey M, McArthur MJ, Montgomery CA Jr, Butel JS, Bradley A, Donehower LA (1993) Spontaneous and carcinogen-induced tumorigenesis in p53-deficient mice. Nature Genet 5: 225–229

Harvey M, Vogel H, Morris D, Bradley A, Bernstein A, Donehower LA (1995) A mutant p53 transgene accelerates tumour development in heterozygous but not nullizygous p53-deficient mice. Nature Genet 9: 305–311

Hayashi K, Schmid W (1975) Tandem duplication of q14 and dicentric formation by end-to-end chromosome fusion in ataxia telangiectasia. Humangenetik 30: 135–141

Heartlein MW, Tsuji H, Latt SA (1987) 5-Bromodeoxyuridine-dependent increase in sister chromatid exchange formation in Bloom's syndrome is associated with reduction in topoisomerase II activity. Exp Cell Res 169: 245–254

Hecht F, Hecht BK (1990) Cancer in ataxia-telangiectasia patients. Cancer Genet Cytogenet 46: 9–19

Hecht F, Koler RD, Rigas DA, Dahnke GS, Case MP, Tisdale V, Miller RW (1966) Leukaemia and lymphocytes in ataxia-telangiectasia. Lancet 2: 1193

Hecht F, McCaw BK, Koler RD (1973) Ataxia telangiectasia: clonal growth of translocation lymphocytes. N Engl J Med 289: 286–291

Heddle JA, Krepinsky AB, Marshall RR (1983) Cellular sensitivity to mutagens and carcinogens in the chromosome-breakage and other cancer-prone syndromes. In: German J (ed) Chromosome mutation and neoplasia. Liss, New York, pp 203–234

Herbst RA, Larson A, Weiss J, Cavanee WK, Hampton GM, Arden KC (1995) A defined region of loss of heterozygosity at 11q23 in cutaneous malignant melanoma. Cancer Res 55: 2494–2496

Higurashi M, Conen PE (1971) In vitro chromosomal radiosensitivity in Fanconi's anemia. Blood 38: 336–342

Hilgers G, Abrahams PJ, Schouten R, Cornelis JJ, Lehmann AR, Van der Eb AJ, Rommelaere J (1987) Cells of patients with ataxia telangiectasia show a normal capacity of radio-induced reactivation of damaged HSV-1 virus. CR Soc Biol (Paris) 181: 432–438

Hilgers G, Abrahams PJ, Chen YQ, Schouten R, Cornelis JJ, Lowe JE, van der Eb AJ, Rommelaere J (1989) Impaired recovery and mutagenic SOS-like responses in ataxia telangiectasia cells. Mutagenesis 4: 271–276

Hill RD (1976) Familial cancer on a Scottish island. Br Med J 2: 401–402

Hoehn H, Kubbies M, Schindler D, Poot M, Rabinovitch PS (1989) BrdU-Hoechst flow cytometry links the cell kinetic defect of Fanconi anemia to oxygen hypersensitivity. In: Schroeder-Kurth TM, Auerbach AD, Obe G (eds) Fanconi anemia. Springer, Berlin Heidelberg New York, pp 161–173

Hojo ET, van Diemen PC, Darroudi F, Natarajan AT (1995) Spontaneous chromosomal aberrations in Fanconi anaemia, ataxia telangiectasia fibroblast and Bloom's syndrome lymphoblastoid cell lines as detected by conventional cytogenetic analysis and fluorescence in situ hybridisation (FISH) technique. Mutat Res 334: 59–69

Huret JL, Tanzer J, Guilhot F, Frocrain-Herchkovitch C, Savage JR (1988) Karyotype evolution in the bone marrow of a patient with Fanconi anemia: breakpoints in clonal anomalies of this disease. Cytogenet Cell Genet 48: 224–227

Iizuka M, Sugiyama Y, Shiraishi M, Jones C, Sekiya T (1995) Allelic losses in human chromosome 11 in lung cancers. Genes Chromosomes Cancer 13: 40–46

Imaly JA, Linn S (1988) DNA damage and oxygen radical toxicity. Science 240: 1302–1309

Ishida R, Buchwald M (1982) Susceptibility of Fanconi's anemia lymphoblasts to DNA-cross-linking and alkylating agents. Cancer Res 42: 4000–4006

Ishizaki K, Yagi T, Inoue M, Nikaido O, Takebe H (1981) DNA repair in Bloom's syndrome fibroblasts after UV irradiation or treatment with mitomycin C. Mutat Res 80: 213–219

Ito I, Yoshimoto M, Iwase T, Watanabe S, Katagiri T, Harada Y, Kasumi F, Yasuda S, Mitomi T, Emi M et al (1995) Association of genetic alterations on chromosome 17 and loss of hormone receptors in breast cancer. Br J Cancer 71: 438–441

Jaspers NG, Gatti RA, Baan C, Linssen PC, Bootsma D (1988) Genetic complementation of analysis of ataxia telangiectasia and Nijmegen breakage syndrome: a survey of 50 patients. Cytogenet Cell Genet 49: 66–73

Jeeves WP, Rainbow AJ (1986) An aberration in gamma-ray-enhanced reactivation of irradiated adenovirus in ataxia telangiectasia fibroblasts. Carcinogenesis 7: 381–387

Jimenez G, Yucel J, Rowley R, Subramani S (1992) The *rad3*$^{+}$ gene of Schizosaccharomyces pombe is involved in multiple checkpoint functions and in DNA repair. Proc Natl Acad Sci USA 89: 4952–4956

Joenje H, Gille JJ (1989) Oxygen metabolism and chromosomal breakage in Fanconi anemia In: Schroeder-Kurth TM, Auerbach AD, Obe G (eds) Fanconi anemia Springer, Berlin Heidelberg New York, pp 174–182

Joenje H, Frants RR, Arwert F, de Bruin GJ, Kostense PJ, van de Kamp JJ, de Koning J, Eriksson AW (1979) Erythrocyte superoxide dismutase deficiency in Fanconi's anaemia established by two independent methods of assay. Scand J Clin Lab Invest 39: 759–764

Joenje H, Arwert F, Eriksson AW, de Koning H, Oostra AB (1981) Oxygen-dependence of chromosomal aberrations in Fanconi's anaemia. Nature 290: 142–143

Joenje H, Lo ten Foe JR, Oostra AB, van Berkel CG, Rooimans MA, Schroeder-Kurth T, Wegner RD, Gille JJ, Buchwald M, Arwert F (1995) Classification of Fanconi anemia patients by complementation analysis: evidence for a fifth genetic subtype. Blood 86: 2156–2160

Johansson E, Niemi KM, Siimes M, Pyrhonen S (1982) Fanconi's anemia. Tumor-like warts, hyperpigmentation associated with deranged keratinocytes, and depressed cell-mediated immunity. Arch Dermatol 118: 249–252

Jones LA, Scott D, Cowan R, Roberts SA (1995) Abnormal radiosensitivity of lymphocytes from breast cancer patients with excessive normal tissue damage after radiotherapy: chromosome aberrations after low dose-rate irradiation. Int J Radiat Biol 67: 519–528

Kaiser TN, Lojewski A, Dougherty C, Juergens L, Sahar E, Latt SA (1982) Flow cytometric characterization of the response of Fanconi's anemia cells to mitomycin C treatment. Cytometry 2: 291–297

Kapp LN, Painter RB (1981) DNA fork displacement rates in human cells. Biochim Biophys Acta 656: 36–39

Karlsen F, Rabbitts PH, Sundresan V, Hagmar B (1994) PCR-RFLP studies on chromosome 3p in formaldehyde-fixed, paraffin-embedded cervical cancer tissues. Int J Cancer 58: 787–792

Kastan MB (1995) Ataxia-telangiectasia – broad implications for a rare disorder. N Engl J Med 333: 662–663

Kastan MB, Zhan Q, el-Deiry WS, Carrier F, Jacks T, Walsh WV, Plunkett BS, Vogelstein B, Fornace AJ Jr (1992) A mammalian cell cycle checkpoint pathway utilizing p53 and GADD45 is defective in ataxia-telangiectasia. Cell 71: 587–597

Kaye J, Smith CA, Hanawalt PC (1980) DNA repair in human cells containing photoadducts of 8-methoxypsoralen or angelicin. Cancer Res 40: 696–702

Keegan KS, Holtzman DA, Plug AW, Brainerd EE, Christenson ER, Bentley EM, Meyn MS, Moss SB, Carr AM, Ashley T, Hoetistra M (1996) The ATR and ATM protein kinases associate with different sites along meiotically pairing chromosomes. Genes Develop (in press)

Keldysh PL, Dragani TA, Fleischman EW, Konstantinova LN, Perevoschikov AG, Pierotti MA, Porta GD, Kopnin PB (1993) 11q deletions in human colorectal carcinomas: cytogenetics and restriction fragment length polymorphism analysis. Genes chromosome Cancer 6: 45–50

Kerlikowske K, Grady D, Barclay J, Sickles EA, Eaton E, Ernster V (1993) Positive predictive value of screening mammography by age and family history of breast cancer. JAMA 270: 2444–2450

Khanna KK, Lavin MF (1993) Ionizing radiation and UV induction of p53 protein by different pathways in ataxia-telangiectasia cells. Oncogene 8: 3307–3312

Kidson C, Chen P, Imray P (1982) Ataxia-telangiectasia heterozygotes: dominant expression of ionizing radiation sensitive mutants. In: Bridges BA, Harnden DG (eds) Ataxia-telangiectasia: a cellular and molecular link between cancer neuropathology, and immune deficiency. Wiley, Chichester, pp 363–372

Kim S, Vollberg TM, Ro JY, Kim M, Sirover MA (1986) O6-methylguanine methyltransferase increases before S phase in normal human cells but does not increase in hypermutable Bloom's syndrome cells. Mutat Res 173: 141–145

Kinsella TJ, Mitchell JB, McPherson S, Russo A, Tietze F (1982) In vitro X-ray sensitivity in ataxia telangiectasia homozygote and heterozygote skin fibroblasts under oxic and hypoxic conditions. Cancer Res 42: 3950–3056

Kobayashi Y, Tycko B, Soreng AL, Sklar J (1991) Transrearrangements between antigen receptor genes in normal human lymphoid tissues and in ataxia telangiectasia. J Immunol 147: 3201–3209

Kohchi C, Noguchi K, Tanabe Y, Mizuno D-I, Soma G-I (1994) Constitutive expression of TNF-α and TNF-β genes in mouse embryo: roles of cytokines as regulator and effector of development. Int J Biochem 26: 111–119

Kohn PH, Kraemer KH, Buchanan JK (1982a) Influence of ataxia telangiectasia gene dosage on bleomycin-induced chromosome breakage and inhibition of replication in human lymphoblastoid cell lines Exp Cell Res 137: 387–395

Kohn PH, Whang-Peng J, Levis WR (1982b) Chromosomal instability in ataxia telangiectasia. Cancer Genet Cytogenet 6: 289–302

Kohno T, Takayama H, Hamaguchi M, Takano H, Yamaguchi N, Tsuda H, Hirohashi S, Vissing H, Shimizu M, Oshimura M et al (1993) Deletion mapping of chromosome 3p in human uterine cervical cancer. Oncogene 8: 1825–1832

Kojis TL, Schreck RR, Gatti RA, Sparkes RS (1989) Tissue specificity of chromosomal rearrangements in ataxia-telangiectasia. Hum Genet 83: 347–352

Kojis TL, Gatti RA, Sparkes RS (1991) The cytogenetics of ataxia telangiectasia. Cancer Genet Cytogenet 56: 143–156

Komatsu K, Okumura Y, Kodama S, Yoshida M, Miller RC (1989) Lack of correlation between rediosensitivity and inhibition of DNA synthesis in hybrids A-T × HeLa. Int J Radiat Biol 56: 863–867

Konstantinova LN, Fleischman EW, Knisch VI, Perevoschikov AG, Kipnon BP (1991) Karyotype peculiarities of human colorectal adenocarcinomas. Hum Genet 86: 491–496

Koreth J, Bethwaite PB, McGee JO (1995) mutation at chromosome 11q23 in human non-familial breast cancer: a microdissection microsatellite analysis. J Pathol 176: 11–18

Korkina LG, Samochatova EV, Maschan AA, Suslova TB, Cheremisina ZP, Afanas'ev IB (1992) Release of active oxygen radicals by leukocytes of Fanconi anemia patients. J. Leukoc Biol 52: 357–362

Krepinksy AB, Heddle JA, German J (1979) Sensitivity of Bloom's syndrome lymphocytes to ethyl methanesulfonate. Hum Genet 50: 151–156

Krepinsky AB, Rainbow AJ, Heddle JA (1980) Studies on the ultraviolet light sensitivity of Bloom's syndrome fibroblasts. Mutat Res 69: 357–368

Krepinsky AB, Heddle JA, German J (1989) Sensitivity of Bloom's syndrome lymphocytes to ethyl methanesulfonate. Hum Genet 50: 151–156

Krontiris TG, Devlin B, Karp DD, Robert NJ, Risch N (1993) An association between the risk of cancer and mutations in the HRAS1 minisatellite locus. N Engl J Med 329: 517–523

Kubbies M, Schindler D, Hoehm H, Schinzel A, Rabinovitch PS (1985) Endogenous blockage and delay of the chromosome cycle despite normal recruitment and growth phase explain poor proliferation and frequent endomitosis in Fanconi's anemia cells. Am J Hum Genet 37: 1022–1030

Kuhn EM (1980) Effects of X-irradiation in G_1 and G_2 on Bloom's syndrome and normal chromosomes. Hum Genet 54: 335–341

Kuhn EM, Therman E (1979) No increased chromosome breakage in three Bloom's syndrome heterozygotes. J Med Genet 16: 219–222

Kuhn EM, Therman E (1986) Cytogenetics of Bloom's syndrome. Cancer Genet Cytogenet 22: 1–18

Kuller LH, Modan B (1992) Risk of breast cancer in ataxia-telangiectasia. N Engl J Med 326: 1357

Kurchgessner CU, Patil CK, Evans JW, Cuomo CA, Fried LM, Carter T, Oettinger MA, Brown JM (1995) DNA-dependent kinase (p350) as a candidate gene for the murine SCID defect. Science 267: 1178–1183

Kurihara T, Inoue M, Tatsumi K (1987a) Hypersensitivity of Bloom's syndrome fibroblasts to N-ethyl-N-nitrosourea. Mutat Res 184: 147–151

Kurihara T, Tatsumi K, Takahashi H, Inoue M (1987b) Sister-Chromatid exchanges induced by ultraviolet light in Bloom's syndrome fibroblasts. Mutat Res 183: 197–202

Kusunoki Y, Hayashi T, Hirai Y, Kushiro J, Tatsumi K, Kurihara T, Zghal M, Kamoun MR, Takebe H, Jeffreys A, et al. (1994) Increased rate of spontaneous mitotic recombination in T lymphocytes from a Bloom's syndrome patient using a flow-cytometric assay at HLA-A locus. Jpn J Cancer Res 85: 610–618

Kyoizumi S, Nakamura N, Takebe H, Tatsumi K, German J, Akiyama M (1989) frequency of variant erythrocytes at the glycophorin-A locus in two Bloom's syndrome patients. Mutat Res 214: 215–222

Lambert MW, Tsongalis GJ, Lambert WC, Hang B, Parrish DD (1992) Defective DNA endonuclease activities in Fanconi's anemia cells, complementation groups A and B. Mutat Res 273: 57–71

Land CE (1992) Risk of breast cancer in ataxia-telangiectasia. N Engl J Med 326: 1359–1361

Landau JW, Sasaki MS, Newcomer VD, Norman A (1966) Bloom's syndrome: the syndrome of telangiectatic erythema and growth retardation. Arch Dermatol 94: 687–694

Lander ES, Botstein D (1987) Homozygosity mapping: a way to map human recessive traits with the DNA of inbred children. Science 236: 1567–1570

Lane DP (1992) Cancer: p53, guardian of the genome. Nature 358: 15–16

Langlois RG, Bigbee WL, Jensen RH, German J (1989) Evidence for increased in vivo mutation and somatic recombination in Bloom's syndrome. Proc Natl Acad Sci USA 86: 670–674

Laquerbe A, Moustacchi E, Fuscoe JC, Papadopoulo D (1995) The molecular mechanism underlying formation of deletions in Fanconi anemia cells may involve a site-specific recombination. Proc Natl Acad Sci USA 92: 831–835

Larrick JW, Wright SC (1990) Cytotoxic mechanism of tumor necrosis factor-α. FASEB J 4: 3215–3223

Latt SA, Stetten G, Juergens LA, Buchanan GR, Gerald PS (1975) Induction by alkylating agents of sister chromatid exchanges and chromatid breaks in Fanconi's anemia. Proc Natl Acad Sci USA 72: 4066–4070

Lavin MF, Le Poidevin P, Bates P (1992) Enhanced levels of radiation-induced G_2 phase delay in ataxia telangiectasia heterozygotes. Cancer Genet Cytogenet 60: 183–187

Lavin MF, Bennett I, Ramsay J, Gardiner RA, Seymour GJ, Farrell A, Walsh M (1994) Identification of a potentially radiosensitive subgroup among patients with breast cancer. J Natl Cancer Inst 86: 1627–1634

Leach FS, Nicolaides NC, Papadopoulos N, Liu B, Jen J, Parsons R et al (1993) Mutations of a mutS homolog in hereditary nonpolyposis colon cancer. Cell 75: 1214–1225

Lees-Miller SP, Godbout R, Chan DW, Weinfeld M, Day III RS, Baron G, Allalunis-Tuner J (1995) Absence of p350 subunit of DNA-activated protein kinase from a radiosensitive human cell line. Science 267: 1183-1185

Lehman AR, Stevens S (1977) The production and repair of double strand breaks in cells from normal humans and from patients with ataxia telangiectasia. Biochim Biophys Acta 474: 49–60

Levanat S, Gorlin RJ, Fallet S, Johnson DR, Fantasia JE, Bale AE (1996) A two-hit model for developmental defects in Gorlin syndrome. Nature Genet 12: 85–87

Li FP, Hecht F, Kaiser-McCaw B, Baranko PV, Potter NU (1981) Ataxia-pancytopenia: syndrome of cerebellar ataxia, hypoplastic anemia, monosomy 7, and acute myelogenous leukemia. Can Genet cytogenet 4: 189–196

Lipkowitz S, Stern MH, Kirsch IR (1990) Hybrid T cell receptor genes formed by interlocus recombination in normal and ataxia-telangiectasis lymphocytes. J Exp Med 172: 409–418

Lipkowitz S, Garry VF, Kirsch IR (1992) Interlocus V-J recombination measures genomic instability in agriculture workers at risk for lymphoid malignancies. Proc Natl Acad Sci USA 89: 5301–5305

Little JB, Nagasawa H (1985) Effect of confluent holding on potentially lethal damage repair, cell cycle progression, and chromosomal aberrations in human normal and ataxia-telangiectasia fibroblasts. Radiat Res 101: 81–93

Little JB, Nove J, Strong LC, Nichols WW (1988) Survival of human diploid skin fibroblasts from normal individuals. Int J Radiat Biol 54: 899–910

Loeb LA (1991) Mutator phenotype may be required for multistage carcinogenesis. Cancer Res 51: 3075–3079

Loeffler JS, Harris JR, Dahlberg WK, Little JB (1990) In vitro radiosensitivity of human diploid fibroblasts derived from women with unusually sensitive clinical responses to definitive radiation therapy for breast cancer. Radiat Res 121: 227–231

Lonn U, Lonn S, Nylen U, Winblad G, German J (1990) An abnormal profile of DNA replication intermediates in Bloom's syndrome. Cancer Res 50: 3141–3145
Louis-Bar D (1941) Sur un syndrome progressif comprenant des télangiectasies capillaires cutanées et conjonctivales symétriques, à disposition naevode et de troubles cérébelleux. Confin Neurol (Basel) 4: 32–42
Lowe SW, Ruley HE, Jacks T, Houseman DE (1993) p53-dependent apoptosis modulates the cytotoxicity of anticancer agents. Cell 74: 957–967
Lu X, Lane DP (1993) Differential induction of transcriptionally active p53 following UV or ionizing radiation: defects in chromosome instability syndromes? Cell 75: 765–778
Lu YY, Jhanwar SC, Cheng JQ, Testa JR (1994) Deletion mapping of the short arm of chromosome 3 in human malignant mesothelioma. Genes Chromosomes Cancer 9: 76–80
Lynch HT, Lynch JF (1994) 25 years of HNPCC. Anticancer Res 14: 1617–1624
Maciejewski JP, Selleri C, Sato T, Anderson S, NS Y (1995) Increased expression of Fas antigen on bone marrow CD34+ cells of patients with aplastic anaemia. Br J Haematol 91: 245–252
Malkin D (1993) p53 and the Li-Fraumeni syndrome. Cancer Genet Cytogenet 66: 83–92
Marshall E (1996) Klausner's unconventional field station in Seattle. Science 271: 1224–1225
Marx MP, Smith S (1989) Significance of cellular sensitivity in a group of parents of Fanconi anemia patients. In: Schroeder-Kurth TM, Auerbach AD, Obe G (eds) Fanconi anemia. Springer, Berlin Heidelberg New York, pp 137–144
Marx MP, Smith S, Heyns AD, van Tonder IZ (1983) Fanconi's anemia: a cytogenetic study on lymphocyte and bone marrow cultures utilizing 1,2:3,4-diepoxybutane. Cancer Genet Cytogenet 9: 51–59
Matthews N, Neale M, Jackson S (1987) Tumor cell killing by TNF inhibited by anaerobic conditions, free radical scavengers and inhibitors of arachidonate metabolism. Immunology 62: 153–155
Mavelli I, Ciriolo MR, Rotilio G, De Sole P, Castorino M, Stabile A (1982) Superoxide dismutase, glutathione peroxidase and catalase in oxidative hemolysis. A study of Fanconi's anemia erythrocytes. Biochem Biophys Res Commun 106: 286–290
McDaniel LD, Schultz RA (1992) Elevated sister chromatid exchange phenotype of Bloom syndrome cells in complemented by human chromosome 15. Proc Natl Acad Sci USA 89: 7968–7972
Meyn MS (1993) High spontaneous intrachromosomal recombination rates in ataxia-telangiectasia. Science 260: 1327–1330
Meyn MS (1995) Ataxia-telangiectasia and cellular responses to DNA damage. Cancer Res 55: 5991–6001
Meyn MS, Bainton J, Herzing LBK (1993) Increased rates of spontaneous intrachromosomal mitotic recombination in Fanconi anemia fibroblast lines. Exp Hematol 21: 717
Meyn MS, Strasfeld L, Allen C (1994) Testing the role of p53 in the expression of genetic instability and apoptosis in ataxia-telangiectasia. Int J Radiat Biol 66: S141–149
Meyn MS, Strasfeld L, Allen C (submitted) p53-mediated apoptosis causes radiosensitivity in ataxia-telangiectasia
Miller AB, Howe GR, Sherman GJ, Lindsay JP, Yaffe MJ, Dinner PJ, Risch HA, Preston DL (1989) Mortality from breast cancer after irradiation during fluoroscopic examinations in patients being treated for tuberculosis. N Eng J Med 321: 1285–1289
Mizutani Y, Bonavida B, Koishihara Y, Akamatsu K, Ohsugi Y, Yoshida O (1995) Sensitization of human renal cell carcinoma cells to cis-diamminedichloroplatinum(II) by anti-interleukin 6 monoclonal antibody or anti-interleukin 6 receptor monoclonal antibody. Cancer Res 55: 590–596
Monnat RJ Jr (1992) Werner syndrome: molecular genetics and mechanistic hypotheses. Exp Gerontol 27: 447–453
Moreira AL, Sampaio EP, Zmuidzinas A, Frindt P, Smith K, Kaplan G (1993) Thalidomide exerts its inhibitory action on tumor necrosis factor α by enhancing mRNA degradation. J Exp Med 199: 1675–1680
Morgan JL, Holcomb TM, Morrissey RW (1968) Radiation reaction in ataxia telangiectasia. Am J Dis Child 116: 557–558
Mori S, Kondo N, Motoyoshi F, Yamaguchi S, Kaneko H, Orii T (1990) Diabetes mellitus in a young man with Bloom's syndrome. Clin Genet 38: 387–390
Mori T, Yanagisawa A, Kato Y, Miura K, Nishihira T, Mori S, Nakamura Y (1994) Accumulation of genetic alterations during esophageal carcinogenesis. Hum Mol Genet 3: 1969–1971
Morrell D, Cromartie E, Swift M (1986) Mortality and cancer incidence in 263 patients with ataxia-telangiectasia. J Natl Cancer Inst 77: 89–92
Morris DJ, Reis A (1994) A YAC contig spanning the nevoid basal cell carcinoma syndrome, Fanconi anaemia group C, and xeroderma pigmentosum group A loci on chromosome 9q. Genomics 23: 23–29

Muleris M, Salmon RJ, Dutrillaux B (1990) Cytogenetics of colorectal adenocarcinomas. Cytogenet Cell Genet 46: 143–156

Mullokandov MR, Kholodilov NG, Atkin NB, Burk RD, Johnson AB, Klinger HP (1996) genomic alterations in cervical carcinoma: losses of chromosome heterozygosity and human papilloma virus tumor status. Cancer Res 56: 197–205

Muriel WJ, Lamb JR, Lehmann AR (1991) UV mutation spectra in cell lines from patients with Cockayne's syndrome and ataxia telangiectasia, using the shuttle vector pZ189. Mutat Res 254: 119–123

Murray AW (1992) Creative blocks: cell-cycle checkpoints and feedback controls. Nature 359: 599–604

Nagafuji K, Shibuya T, Harada M, Mizuno S, Takenaka K, Miyamoto T, Okamura T, Gondo H, Niho Y (1995) Functional expression of Fas antigen (CD95) on hematopoietic progenitor cells. Blood 86: 883–889

Nagasawa H, Little JB (1983) Suppression of cytotoxic effect of mitomycin-C by superoxide dismutase in Fanconi's anemia and dyskeratosis congenita fibroblasts. Carcinogenesis 4: 795–799

Nagasawa H, Latt SA, Lalande ME, Little JB (1985) Effects of X-irradiation on cell-cycle progression, induction of chromosomal aberrations and cell killing in ataxia telangiectasia (AT) fibroblasts. Mutat Res 148: 71–82

Nagasawa H, Kraemer KH, Shiloh Y, Little JB (1987) Detection of ataxia telangiectasia heterozygous cell lines by postirradiation cumulative labelling index: measurements with coded samples. Cancer Res 47: 398–402

Natarajan AT, Meijers M, van Zeeland AA, Simons JW (1982) Attempts to detect ataxia telangiectasia (AT) heterozygotes by cytogenetical techniques. Cytogenet Cell Genet 33: 145–151

National Research Council (1990) Health effects of exposure to low levels of ionizing radiation. In: "BEIR V". National Academy Press, Washington DC

Negrini M, Rasio D, Hampton GM, Sabbiloni S, Rattan S, Carter SL, Rosenberg AL, Schwartz GF, Shiloh Y, Cavenee WK, Croce CM (1995) Definition and refinement of chromosome 11 regions of loss of heterozygosity in breast cancer: identification of a new region at 11q23.3. Cancer Res 55: 3003–3007

Nelson WG, Kastan MB (1994) DNA strand breaks: the DNA template alterations that trigger p53-dependent DNA damage response pathways. Mol Cell Biol 14: 1815–1823

Nicolaides NC, Papadopoulos N, Lie B, Wei Y-F, Carter KC, Ruben SM et al (1994) Mutations of two homologous in hereditary non-polyposis colon cancer. Nature 371: 75–80

Nicotera TM, Notaro J, Notaro S, Schumer J, Sandberg AA (1989) Elevated superoxide dismutase in Bloom's syndrome: a genetic condition of oxidative stress. Cancer Res 49: 5239–5243

Nordenson I (1977) Effect of superoxide dismutase and catalase on spontaneously occurring chromosome breaks in patients with Fanconi's anemia. Hereditas 86: 147–150

Norman A, Withers HR (1993) Mammography screening for A-T heterozygotes. In: Gatti RA, painter RB (eds) Ataxia-telangiectasia. Springer, Berlin Heidelberg New York, pp 137–140

Norman A, Kagan AR, Chan SL (1988) The importance of genetics for the optimization of radiation therapy. Am J Clin Oncol 11: 84–88

Nowell PC (1976) The clonal evolution of tumor cell populations. Science 194: 23–28

Ockey CH, Saffhill R (1986) Delayed DNA maturation, a possible cause of the elevated sister-chromatid exchange in Bloom's syndrome. Carcinogenesis 7: 53–57

Ogasawara S, Maesawa C, Tamura G, Satodate R (1995) Frequent microsatellite alterations on chromosome 3p in esophageal squamous cell carcinoma. Cancer Res 55: 891–894

Okahata S, Kobayashi Y, Usui T (1980) Erythrocyte superoxide dismutase activity in Fanconi's anaemia. Clin Sci 58: 173–175

Oto S, Miyamoto S, Kudoh F, Horie H, Kinugawa N, Okimoto Y (1992) Treatment for B-cell-type lymphoma in a girl associated with Bloom's syndrome. Clin Genet 41: 46–50

Oxford JM, Harnden DG, Parrington JM, Delhanty JD (1975) Specific chromosome aberrations in ataxia telangiectasia. J Med Genet 12: 251–262

Painter RB, Young BR (1980) Radiosensitivity in ataxia-telangiectasia: a new explanation. Proc Natl Acad Sci USA 77: 7215–7217

Pandita TK, Hittelman WN (1992a) The contribution of DNA and chromosome repair deficiencies to the radiosensitivity of ataxia-telangiectasia. Radiat Res 131: 214–223

Pandita TK, Hittelman WN (1992b) Initial chromosome damage but not DNA damage is greater in ataxia telangiectasia cells. Radiat Res 130: 94–103

Pandita TK, Hittelman WN (1994) Increased levels of chromosome damage and heterogeneous chromosome repair in ataxia telangiectasia heterozygote cells. Mutat Res 310: 1–13

Pandita TK, Hittelman WN (1995) Evidence of a chromatin basis for increased mutagen sensitivity associated with multiple primary malignancies of the head and neck. Int J Cancer 61: 738–743

Pandita TK, Pathak S, Geard CR (1995) Chromosome end associations, telomeres and telomerase activity in ataxia telangiectasia cells. Cytogenet Cell Genet 71: 86–93

Papadopoulo D, Guillouf C, Mohrenweiser H, Moustacchi E (1990a) Hypomutability in Fanconi anaemia cells is associated with increased deletion frequency at the HPRT locus. Proc Natl Acad Sci USA 87: 8383–8387

Papadopoulo D, Porfirio B, Moustacchi E (1990b) Mutagenic response of Fanconi's anemia cells from a defined complementation group after treatment with photoactivated bifunctional psoralens. Cancer Res 50: 3289–3294

Papadopoulos N, Nicolaides N, Wei Y-F, Ruben SM, Carter KC, Rosen CA et al (1994) Mutation of a mutL homolog in hereditary colon cancer. Science 263: 1625–1629

Parshad R, Sanford KK, Jones GM (1983) Chromatid damage after G_2 phase X-irradiation of cells from cancer-prone individuals implicates deficiency in DNA repair. Proc Natl Acad Sci USA 88: 7615–7619

Parshad R, Sanford KK, Jones GM, Tarone RE (1985) G_2 chromosomal radiosensitivity of ataxia-telangiectasia heterozygotes. Cancer Genet Cytogenet 14: 163–168

Partridge M, Kiguwa S, Langdon JD (1994) Frequent deletion of chromosome 3p in oral squamous cell carcinoma. Eur J Cancer B Oral Oncol 30B: 248–251

Passarge E (1991) Bloom's syndrome: the German experience. Ann Genet 34: 179–197

Patchen ML, Mac Vittie TJ, Williams JL, Schwartz GN, Souza LM (1991) Administration of interleukin-6 stimulates multilineage hematopoiesis and accelerates recovery from radiation-induced hematopoietic depression. Blood 77: 472–480

Paterson MC, Anderson AK, Smith BP, Smith PJ (1979) Enhanced radiosensitivity of cultured fibroblasts from ataxia telangiectasia heterozygotes manifested by defective colony-forming ability and reduced DNA repair replication after hypoxic γ-irradiation. Cancer Res 39: 3725–3734

Paterson MC, Bech-Hansen NT, Smith PJ, Mulvihill JJ (1984) Radiogenic neoplasia, cellular radiosensitivity and faulty DNA repair. In: Boice J, Frauni JF (eds) Radiation carcinogenesis: epidemiology and biological significance. Raven, New York, pp 319–336

Paterson MC, MacFarlane SJ, Gentner NE, Smith BP (1985) Cellular hypersensitivity to chronic γ-radiation in cultured fibroblasts from ataxia-telangiectasia heterozygotes. In: Gatti RA, Swift M (eds) Ataxia-telangiectasia: genetics, neuropathology, and immunology of a degenerative disease of childhood. Liss, New York, pp 73–87

Pavlakis SG, Frissora CL, Giampietro PF, Davis JG, Gould RJ, Adler-Brecher B, Auerbach AD (1992) Fanconi anemia: a model for genetic causes of abnormal brain development. Dev Med Child Neurol 34: 1081–1084

Pejovic T (1995) Genetic changes in ovarian cancer. Ann Med 27: 73–78

Perdahl EB, Naprstek BL, Wallace WC, Lipton JM (1994) Erythroid failure in Diamond-Blackfan anemia is characterized by apoptosis. Blood 83: 645–650

Perkins J, Timson J, Emery AE (1969) Clinical and chromosome studies in Fanconi's aplastic anaemia. J Med Genet 6: 28–33

Peterson RDA, Good RA (1968) Ataxia-telangiectasia. In: Bergsma D, Good RA (eds) Birth defects – immunologic defiency disease in man. National Foundation, March of Dimes, New York, pp 370–377

Peterson RDA, Funkhouser JD, Tuck-Muller CM, Gatti RA (1992) Cancer susceptibility in ataxia-telangiectasia. Leukemia 6 [Suppl 1]: 8–13

Petridou M, Barrett AJ (1990) Physical and laboratory characteristics of heterozygote carriers of the Fanconi aplasia gene. Acta Paediatr Scand 79: 1069–1074

Pfeiffer RA (1970) Chromosomal abnormalities in ataxia-telangiectasia (Louis Bar's syndrome). Humangenetik 8: 302–306

Phipps J, Nasim A, Miller DR (1985) Recovery, repair and mutagenesis in Schizosaccharomyces pombe. Adv Genet 23: 1–72

Pincheira J, Bravo M, Lopez-Saez JF (1988) Fanconi's anemia lymphocytes: effect of caffeine, adenosine and niacinamide during G_2 prophase. Mutat Res 199: 159–165

Pippard EC, Hall AJ, Barker DJ, Bridges BA (1988) Cancer in homozygotes and heterozygotes of ataxia-telangiectasia and xeroderma pigmentosum in Britain. Cancer Res 48: 2929–2932

Poll EH, Abrahams PJ, Arwert F, Eriksson AW (1984) Host-cell reactivation of cis-diamminedichloroplatinum(II)-treated SV40 DNA in normal human, Fanconi anaemia and xeroderma pigmentosum fibroblasts. Mutat Res 132: 181–187

Poll EHA, Arwert F, Joenje H, Wanamarta AH (1985) Differential sensitivity of Fanconi anaemia lymphocyotes to the clastogenic action of cis-diamminedichloroplatinum (II) and trans-diamminedichloroplatinum (II). Hum Genet 71: 206–210

Poot M, Verkerk A, Koster JF, Jongkind JF (1986) De novo synthesis of glutathione in human fibroblasts during in vitro ageing and in some metabolic diseases as measured by a flow cytometric method. Biochim Biophys Acta 883: 580–584

Porfirio B, Dallapiccola B, Mokini V, Alimena G, Gandini E (1983) Failure of diepoxybutane to enhance sister chromatid exchange levels in Fanconi's anemia patients and heterozygotes. Hum Genet 63: 117–120

Potter NU, Sarmousakis C, Li FP (1983) Cancer in relatives of patients with aplastic anemia. Cancer Genet Cytogenet 9: 61–69

Pritsos CA, Sartorelli AC (1986) Generation of reactive oxygen radicals through biactivation of mitomycin antibiotics. Cancer Res 46: 3528–3532

Pronk JC, Gibson RA, Savoia A, Wijker M, Morgan NV, Melchionda S, Ford D, Temtamy S, Ortega JJ, Jansen S et al (1995) Localisation of the Fanconi anaemia complementation group A gene to chromosome 16q24.3. Nature Genet 11: 338–340

Rabbitts TH (1991) Translocations, master genes and differences between the origin of acute and chronic leukaemias. Cell 67: 641–644

Ramsay J, Birrell G (1995) Normal tissue radiosensitivity in breast cancer patients. Int J Radiat Oncol Biol Phys 31: 339–344

Randal J (1995a) Are mammograms a good idea for AT gene carriers? J Natl Cancer Inst 87: 1351

Randal J (1995b) ATM gene discovery may quiet carrier cancer risk debate. J Natl Cancer Inst 87: 1350–1351

Rasio T, Negrini M, Manenti G, Dragani TA, Croce CM (1995) Loss of heterozygosity at chromosome 11q in lung adenocarcinoma: identification of three independent regions. Cancer Res 55: 3988–3991

Reed WB, Epstein WL, Boder E, Sedgwick R (1966) Cutaneous manifestations of ataxia-telangiectasia. JAMA 195: 746–753

Rennert G (1991) The value of mammography in different ethnic groups in Israel – analysis of mortality reduction and costs using CAN*Trol. Cancer Detect Prevent 15: 477–481

Rey JP, Scott R, Muller H (1994) Apoptosis is not involved in the hypersensitivity of Fanconi anemia cells to mitomycin C. Cancer Genet Cytogenet 75: 67 71

Riley E, Caldwell R, Swift M (1979) Comparison of clinical features in Fanconi anemia probands and their subsequently diagnosed siblings. Am J Hum Genet 31: 82A

Rosendorff J, Bernstein R (1988) Fanconi's anemia – chromosome breakage studies in homozygotes and hetrozygotes. Cancer Genet Cytogenet 33: 175–183

Rosendorff J, Bernstein R, Macdougall L, Jenkins T (1987) Fanconi anemia: another disease of unusually high prevalence in the Afrikaans population of South Africa. Am J Med Genet 27: 793–797

Rosin MP, German J (1985) Evidence for chromosome instability in vivo in Bloom syndrome: increased numbers of micronuclei in exfoliated cells. Hum Genet 71: 187–191

Rosin MP, Ochs HD, Gatti RA, Boder E (1989) Heterogeneity of chromosomal breakage levels in epithelial tissue of ataxia-telangiectasia homozygotes and heterozygotes. Hum Genet 83: 133–138

Rosselli F, Sanceau J, Wietzerbin J, Moustacchi E (1992) Abnormal lymphokine production: a novel feature of the genetic disease Fanconi anemia. I. Involvement of interleukin-6. Hum Genet 89: 42–48

Rosselli F, Sanceau J, Gluckman E, Wietzerbin J, Moustacchi E (1994) Abnormal lymphokine production: a novel feature of the genetic disease Fanconi anemia. II. In vitro and in vivo spontaneous overproduction of tumor necrosis factor alpha. Blood 83: 1216–1225

Rosselli F, Ridet A, Soussi T, Duchaud E, Alapetite C, Moustacchi E (1995) p53-dependent pathway of radio-induced apoptosis is altered in Fanconi anemia. Oncogene 10: 9–17

Rothstein JL, Johnson D, DeLoia JA, Skowronski J, Solter D, Knowles B (1992) Gene expression during preimplantation mouse development. Genes Dev 6: 1190–1201

Rousset S, Nocentini S, Revet B, Moustacchi E (1990) Molecular analysis by electron microscopy of the removal of psoralen-photoinduced DNA cross-links in normal and Fanconi's anemia fibroblasts. Cancer Res 50: 2443–2448

Rowell S, Newman B, Boyd J, King M-C (1994) Inherited predisposition to breast and ovarian cancer. Am J Hum Genet 55: 861–865

Rudolph NS, Latt SA (1989) Flow cytometric analysis of X-ray sensitivity in ataxia telangiectasia.Mutat Res 211: 31–41

Rudolph NS, Nagasawa H, Little JB, Latt SA (1989) Identification of ataxia telangiectasia heterozygotes by flow cytometric analysis of X-ray damage. Mutat Res 211: 19–29

Russell KJ, Wiens LW, Demers GW, Galloway DA, Plon SE, Groudine M (1995) Abrogation of the G_2 checkpoint results in differential radiosensitization of the G_1 checkpoint-deficient and G_1 checkpoint-competent cells, Cancer Res 55: 139–142

Saadi AA, Palutke M, Kumar GK (1980) Evolution of chromosomal abnormalities in sequential cytogenetic studies of ataxia telangiectasia. Hum Genet 55: 23–29

Saito H, Hammond AT, Moses RE (1993) Hypersensitivity to oxygen is a uniform and secondary defect in Fanconi anemia cells. Mutat Res 294: 255–262

Sala-Trepat M, Boyse J, Richard P, Papadopoulo D, Moustacchi E (1993) Frequencies of HPRT-lymphocytes and glycophorin A variants erythrocytes in Fanconi anemia patients, their parents and control donors. Mutat Res 289: 115–126

Sanchez Y, Desany BA, Jones WJ, Liu Q, Wang B, Elledge SJ (1996) Regulation of RAD53 by the ATM-like Kinases MEC1 and TEL1 in yeast cell checkpoint pathways. Science 271: 357–360

Sankaranarayanan K, Charkraborty R (1995) Cancer predisposition, radiosensitivity and the risk of radiation-induced cancers, I. Background Radiat Res 143: 121–143

Sasaki MS (1975) Is Fanconi's anaemia defective in a process essential to the repair of DNA cross links? Nature 257: 501-503

Sasaki MS (1978) Fanconi's anemia. A condition possibly associated with a defective DNA repair. In: Hanawalt PC, Friedberg EC, Fox CF (eds) DNA repair mechanisms. Academic, New York, pp 675–684

Sasaki MS (1989) The Japan Society of Human Genetics award lecture. Cytogenetic aspects of cancer – predisposing genes. Jinrui Idengaku Zasshi 34: 1–16

Sasaki MS, Tonomura A (1973) A high susceptibility of Fanconi's anemia to chromosome breakage by DNA cross-linking agents. Cancer Res 33: 1829–1836

Savitsky K, Bar-Shira A, Gilad S, Rotman G, Ziv Y, Vanagaite L, Tagle DA, Smith S, Uziel T, Sfez S et al (1995) A single ataxia telangiectasia gene with a product similar to PI-3 kinase. Science 268: 1749–1753

Savoia A, Centra M, Ianzano L, de Cillis GP, Zelante L, Buchwald M (1995) Characterization of the 5′ region of the Fanconi anaemia group C (FACC) gene. Hum Mol Genet 4: 1321–1326

Scarpa M, Rigo A, Momo F, Isacchi G, Novelli G, Dallapiccola B (1985) Increased rate of superoxide ion generation in Fanconi anemia erythrocytes. Biochem Biophys Res Commun 130: 127–132

Schalch DS, McFarlin DE, Barlow MH (1970) An unusual form of diabetes mellitus in ataxia telangiectasia. N Eng J Med 282: 1369–1402

Schindler D, Hoehn H (1988) Fanconi anemia mutation causes cellular susceptibility to ambient oxygen. Am J Hum Genet 43: 429–435

Schofield D, West DC, Anthony DC, Marshal R, Sklar J (1995) Correlation of loss of heterozygosity at chromosome 9q with histological subtype in medulloblastomas. Am J Pathol 146: 472–480

Schroeder TM (1982) Genetically determined chromosome instability syndromes. Cytogenet Cell Genet 33: 119–132

Schroeder TM, Kurth R (1971) Spontaneous chromosomal breakage and high incidence of leukemia in inherited disease. Blood 37: 96–112

Schroeder TM, Anschütz F, Knopp A (1964) Spontane Chromosomenaberrationen bei familiärer Panmyelopathie Humangenetik 1: 194–196

Schroeder TM, Tilgen D, Kruger J, Vogel F (1976) Formal genetics of Fanconi's anemia. Hum Genet 32: 257–288

Schultz JC, Shahidi NT (1993) Tumor necrosis factor-alpha overproduction in Fanconi's anemia. Am J Hematol 42: 196–201

Schwarz A, Bhardwaj R, Aragane Y, Nahnke K, Riemann H, Metze D, Luger TA, Schwartz T (1995) Ultraviolet-B-induced apoptosis of keratinocytes: evidence for partial involvement of tumor necrosis factor-α in the formation of sunburn cells. J Invest Dermatol 104: 922–927

Scott D, Jones LA, Elyan SAG, Spreadborough A, Cowan R, Ribiero G (1993) Identification of A-T heterozygotes. In: Gatti RA, Painter RB (eds) Ataxia-telangiectasia. Springer, Berlin Heidelberg New York, pp 101–116

Scott D, Spreadborough AR, Roberts SA (1994) Radiation-induced G_2 delay and spontaneous chromosome aberrations in ataxia-telangiectasia homozygotes and heterozygotes. Int J Radiat Biol 66: S157–163

Seal G, Brech K, Karp SJ, Cool BL, Sirover MA (1988) Immunological lesions in human uracil DNA glycosylase: association with Bloom syndrome. Proc Natl Acad Sci USA 85: 2339–2343

Sedgwick RP, Boder E (1991) Ataxia-telangiectasia. In: deJong JMBV (ed) Hereditary neuropathies and spinocerebellar atrophies. Elsevier Science, New York, pp 347–423

Selleri C, Satao T, Anderson S, Young NS, Maciejewski JP (1995) Interferon-gamma and tumor necrosis factor-alpha suppress both early and late stages of hematopoiesis and induce programmed cell death. J Cell Phys 165: 538–546

Seres DS, Fornace AJ Jr (1982) Normal response of Fanconi's anemia cells to high concentrations of O_2 as determined by alkaline elution. Biochim Biophys Acta 698: 237–242

Seyschab H, Bretzel G, Friedl R, Schindler D, Sun Y, Hoehn H (1994) Modulation of the spontaneous G_2 phase blockage in Fanconi anemia cells by caffeine: differences from cells arrested by X-irradiation. Mutat Res 308: 149–157

Seyschab H, Friedl R, Sun Y, Schindler D, Hoehn H, Hentze S, Schroeder-Kurth T (1995) Comparative evaluation of diepoxybutane sensitivity and cell cycle blockage in the diagnosis of Fanconi anemia. Blood 85: 2233–2237

Shanley SM, Dawkins H, Wainwright BJ, Wicking C, Heenan P, Eldon M, Searle J, Chenevix-Trench G (1995) Fine deletion mapping on the long arm of chromosome 9 in sporadic and familiar basal cell carcinomas. Hum Mol Genet 4: 129–133

Shiloh Y, Tabor E, Becker Y (1982a) Abnormal response of ataxia-telangiectasia cells to agents that break the deoxyribose moiety of DNA via a targeted free radical mechanism. Carcinogenesis 4: 1317–1322

Shiloh Y, Tabor E, Becker Y (1982b) The response of ataxia-telangiectasia homozygous and heterozygous skin fibroblasts to neocarzinostatin. Carcinogenesis 3: 815–820

Shiloh Y, Parshad R, Sanford KK, Jones GM (1986) Carrier detection in ataxia-telangiectasia. Lancet 1: 689–690

Shiraishi Y, Sandberg AA (1978) Effects of mitomycin C on sister chromatid exchange in normal and Bloom's syndrome fibroblast. Mutat Res 49: 233–238

Shiraishi Y, Taguchi T, Ozawa M, Bamazai R (1989) Different mutations responsible for the elevated sister-chromatid exchange frequencies in Bloom syndrome and X-irradiated B-lymphoblastoid cell lines originating from acute leukemia. Mutat Res 211: 273–278

Skirnisdottir S, Eiriksdottir G, Baldursson T, Barkarsdottir RB, Egilsson V, Ingvarrson S (1995) High frequency of allelic imbalance at chromosome region 16q22–23 in human breast cancer: correlation with high PgR and low S phase. Int J Cancer 64: 112–116

Spector BD, Filipovick AH, Perry GS III, Kersey JH (1982) Epidemiology of cancer in ataxia-telangiectasia. In: Bridges BA, Harnden DG (eds) Ataxia-telangiectasia. Wiley, New York, pp 103–138

Sreekantaiah C, DeBraekeleer M, Haas O (1991) Cytogenetic findings in cervical carcinoma: a statistical approach. Cancer Genet Cytogenet 53: 75–81

Stacey M, Thacker S, Taylor AM (1989) Cultured skin keratinocytes from both normal individuals and basal cell naevus syndrome patients are more resistant to gamma-rays and UV light compared with cultured skin fibroblasts. Int J Radiat Biol 56: 45–58

Stark R, Andre C, Thierry D, Cherel M, Galibert F, Gluckman E (1993) The expression of cytokine and cytokine receptor genes in long-term bone marrow culture in congenital and acquired bone marrow hypoplasias. Br J Haematol 83: 560–566

Strathdee CA, Duncan AM, Buchwald M (1992a) Evidence for at least four Fanconi anaemia genes including FACC on chromosome 9. Nature Genet 1: 196–198

Strathdee CA, Gavish H, Shannon WR, Buchwald M (1992b) Cloning of cDNAs for Fanconi's anaemia by functional complementation. Nature 358: 434

Strich S (1963) Pathological findings in 3 cases of ataxia-telangiectasia. Acta Neurol Scand 42: 354–366

Strober W, Wochner RD, Barlow MH, McFarlin DE, Waldmann TA (1968) Immunoglobulin metabolism in ataxia telangiectasia. J Clin Invest 47: 1905–1915

Sullaba L, Henner K (1926) Contribution a l'indépendance de l'athétose double idiopathique et congénitale. Atteinte familiale, syndrome dystrophique, signe du réseau vasculaire conjonctival, intégrité psychique. Rev Neurol 1: 541–562

Summers WC, Sarkar SN, Glazer PM (1985) Direct and inducible mutagenesis in mammalian cells. Cancer Surv 4: 517–528

Sung P, Prakash L, Matson SW, Prakash S (1987) RAD3 protein of Saccharomyces cerevisiae is a DNA helicase. Proc Natl Acad Sci USA 84: 8951–8955

Swift M (1971) Fanconi's anaemia in the genetics of neoplasia. Nature 230: 370–373

Swift M (1994) Ionizing radiation, breast cancer, and ataxia-telangiectasia. J Natl Cancer Inst 86: 1571–1572

Swift M, Sholman L, Perry M, Chase C (1976) Malignant neoplasms in the families of patients with ataxia-telangiectasia. Cancer Res 35: 208–215

Swift M, Caldwell RJ, Chase C (1980) Reassessment of cancer predisposition of Fanconi anaemia heterozygotes. J Natl Cancer Inst 65: 863–867

Swift M, Morrell D, Cromartie E, Chamberlin AR, Skolnick MH, Bishop DT (1986) The incidence and gene frequency of ataxia-telangiectasia in the United States. Am J Hum Genet 39: 573–583
Swift M, Reitnauer PJ, Morrel D, Chase CL (1987) Breast and other cancers in families with ataxia-teloangeictasia. N Engl J Med 316: 1289–1294
Swift M, Morrell D, Massey RB, Chase CL (1991) Incidence of cancer in 161 families affected by ataxia-telangiectasia. N Engl J Med 325: 1831–1836
Szalay GC (1963) Dwarfism with skin manifestations. J Pediat 62: 686–695
Taccioli GE, Gottlieb TM, Blunt T, Priestley A, Demengeot J, Mizuta R, Lehmann AR, Alt FW, Jackson SP, Jeggo PA (1994) Ku80: product of the XRCC5 gene and its role in DNA repair and V(D)J recombination. Science 265: 1442–1445
Taylor AMR (1978) Unrepaired DNA strand breaks in irradiated ataxia-telangiectasia lymphocytes suggested from cytogenetic observations. Mutat Res 50: 407–418
Taylor AMR (1992) Ataxia telangiectasia genes and predisposition to leukaemia, lymphoma and breast cancer. Br J Cancer 66: 5–9
Taylor AMR, Harnden DG, Arlett CF, Harcourt SA, Lehmann AR, Stevens S, Bridges BA (1975) Ataxia telangiectasia: a human mutation with abnormal radiation sensitivity. Nature 258: 427–428
Taylor AMR, Metcalfe JA, McConville C (1989) Increased radiosensitivity and the basic defect in ataxia telangiectasia. Int J Radiat Biol 56: 677–684
Taylor AMR, Oxford JM, Metcalfe JA (1981) Spontaneous cytogenetic abnormalities in lymphocytes from thirteen patients with ataxia telangiectasia. Int J Cancer 27: 311–319
Teebor GW, Duker NJ (1975) Human endonuclease activity for DNA apurinic sites. Nature 158: 544–547
Thiberville L, Bourguignon J, Metayer J, Bost F, Diarra-Mehrpour M, Bignon J, Lam S, Martin JP, Nouvet G (1995) Frequency and prognostic evaluation of 3p21–22 allelic losses in non-small-cell lung cancer. Int J Cancer 64: 371–377
Tomlinson IP, Strickland JE, Lee AS, Bromley L, Evans MF, Morton J, McGee JO (1995) Loss of heterozygosity on chromosome 11 q in breast cancer. J Clin Pathol 48: 424–428
Tomlinson IPM, Gammack AJ, Stickland JE, Mann GJ, MacKier RTM, Iefford RF, McGee JO (1993) Loss of heterozygosity in malignant melanoma at loci on chromosome 11 and 17 implicated in the pathogenesis of other cancers. Genes Chromosomes Cancer 7: 169–172
Troelstra C, Jaspers NG (1994) Recombination and repair. Ku starts at the end. Curr Biol 4: 1149–1151
Tsai-Pflugfelder M, Liu L, Liu A, Tewey K, Whang-Peng J, Knutsen T, Huebner K, Croce C, Wang J (1988) Cloning and sequencing of cDNA encoding human DNA topoisomerase II and localizating of the gene to chromosome region 17q21–22. Proc Natl Acad Sci USA 85: 7177–7181
Turesson I (1990) Individual variation and dose dependency in the progression rate of skin telangiectasias. Int J Radiat Oncol Biol Phys 19: 1569–1574
Uhrhammer N, Cancannon P, Huo Y, Nakamura Y, Gatti RA (1994) A pulsed-field gel electrophoresis map in the ataxia-telangiectasia region of chromosome 11q22.3. Genomics 20: 278–280
Uhrhammer N, Lange E, Porras O, Naeim A, Chen X, Sheikhavandi S, Chiplunkar S, Yang L, Dandekar S, Liang T et al (1995) Sublocalization of an ataxia-telangiectasia gene distal to D11S384 by ancestral haplotyping in Costa Rican families. Am J Hum Genet 57: 103–111
van Laar T, Steegenga WT, Jochemsen AG, Terleth C, van der Eb AJ (1994) Bloom's syndrome cells GM1492 lack detectable p53 protein but exhibit normal G_1 cell-cycle arrest after UV irradiation. Oncogene 9: 981–983
van Leeuven H (1933) Ein Fall von "konstitutioneller infantiler perniziösähnlicher Anämie" (Fanconi). Folia Haematol (Lpz) 49: 434–443
Verlander PC, Lin JD, Udono MU, Zhang Q, Gibson RA, Mathew CG, Auerbach AD (1994) Mutation analysis of the Fanconi anemia gene FACC. Am J Hum Genet 54: 595–601
Verlander PC, Kaporis A, Liu Q, Zhang Q, Seligsohn U, Auerbach AD (1995) Carrier frequency of the IVS4 + 4 A $\rightarrow$ T mutation of the Fanconi anemia gene FACC in the Ashkenazi Jewish population. Blood 86: 4034–4038
Vijayalakshmi E, Wunder P, Schroeder TM (1985) Spontaneous 6-thioguanine-resistant lymphocytes in Fanconi anemia patients and their heterozygous parents. Hum Genet 70: 264–270
Vogel VG (1994) Screening younger women at risk for breast cancer. Monogr Natl Cancer Inst 16: 55–60
Vogelstein B, Kinzler KW (1993) The multistep nature of cancer. Trends Genet 9: 138–141
Vollberg TM, Seal G, Sirover MA (1987) Monoclonal antibodies detect conformational abnormality of uracil DNA glycosylase in Bloom's syndrome cells. Carcinogenesis 8: 1725–1729
Waghray M, Sigut D, Einspenner M, Kunhi M, al-Sedairy ST, Hannan MA (1992) Chronic gamma-irradiation results in increased cell killing and chromosomal aberration with specific breakpoints in fibroblast cell strains derived from non-Hodgkin's lymphoma patients. Mutat Res 284: 223–231

Watts PM, Louis EJ, Borts RH, Hickson ID (1995) Sgs 1: a eukaryotic homolog of *E. coli RecQ* that interacts with topoisomerase II in vivo and is required for faithful chromosome segregation. Cell 81: 253–260
Weeks DE, Paterson MC, Lange K, Andrais B, Davis RC, Yoder F, Gatti RA (1991) Assessment of chronic gamma radiosensitivity as an in vitro assay for heterozygote identification of ataxia-telangiectasia. Radiat Res 128: 90–99
Weemaes CM, Bakkeren JA, Haraldsson A, Smeets DF (1991) Immunological studies in Bloom's syndrome. A follow-up report. Ann Genet 34: 201–205
Weemaes CM, Smeets DF, van der Burgt CJ (1994) Nijmegen breakage syndrome: a progress report. Int J Radiat Biol 66: S185–188
Weichselbaum RR, Nove J, Little JB, (1980) X-ray sensitivity of fifty-three human diploid fibroblast cell strains from patients with characterized genetic disorders. Cancer Res 40: 920–925
Weinert TA, Kiser GL, Hartwell LH (1994) Mitotic checkpoint genes in budding yeast and the dependence of mitosis on DNA replication and repair. Genes Dev 8: 652–665
Weitzel JN, Patel J, Smith DM, Goodman A, Safaii H, Ball HG (1994) Molecular genetic changes associated with ovarian cancer. Gynecol Oncol 55: 245–252
Weksberg R, Buchwald M, Sargent P, Thompson MW, Siminovitch L (1979) Specific cellular defects in patients with Fanconi anemia. J Cell Physiol 101: 311–323
Weksberg R, Smith C, Anson-Cartwright L, Maloney K (1988) Bloom syndrome: a single complementation group defines patients of diverse ethnic origin. Am J Hum Genet 42: 816–824
Weshimer K, Swift M (1982) Congenital malformations and developmental disabilities in ataxia-telangiectasia, Fanconi anemia and xeroderma pigmentosum. Am J Hum Genet 34: 781–793
West CM, Hendry J, Scott D et al (1991) 25th Paterson symposium – is there a future for radiosensitivity testing? Br J Cancer 24: 146s–152s
West CM, Elyan SA, Berry P, Cowan R, Scott D (1995) A comparison of the radiosensitivity of lymphocytes from normal donors, cancer patients, individuals with ataxia-telangiectasia (A-T) and A-T heterozygotes. Int J Radiat Biol 68: 197–203
Wevrick R, Clarke CA, Buchwald M (1993) Cloning and analysis of murine Fanconi anemia group C cDNA. Hum Mol Genet 2: 655–662
Whitney M, Thayer M, Reifsteck C, Olson S, Smith L, Jakobs PM, Leach R, Naylor S, Joenje H, Grompe M (1995) Microcell mediated chromosome transfer maps the Fanconi anaemia group D gene to chromosome 3p. Nature Genet 11: 341–343
Wiencke JK, Wara DW, Little JB, Kelsey KT (1992) Heterogeneity in the clastogenic response to X-rays in lymphocytes from ataxia-telangiectasia heterozygotes and controls. Cancer Causes Control 3: 237–245
Willis AE, Lindahl T (1987) DNA ligase I deficiency in Bloom's syndrome. Nature 325: 355–357
Winqvist R, Hampton GM, Mannermaa A, Blanco G, Alavaikko M, Kiviniemi H, Taskinen PJ, Evans GA, Wright FA, Newsham I et al (1995) Loss of heterozygosity for chromosome 11 in primary human breast tumors is associated with poor survival after metastasis. Cancer Res 55: 2660–2664
Woods CG, Taylor AM (1992) Ataxia telangiectasia in the British Isles: the clinical and laboratory features of 70 affected individuals. Q J Med 82: 169–179
Woods CG, Bundey SE, Taylor AMR (1990) Unusual features in the inheritance of ataxia telangiectasia. Hum Genet 84: 555–562
Wooster R, Ford D, Mangion J, Ponder BA, Peto J, Easton DF, Stratton MR (1993) Absence of linkage to the ataxia telangiectasia locus in familial breast cancer. Hum Genet 92: 91–94
Wride MA, Sanders EJ (1995) Potential roles for tumor necrosis factor α during embryonic development. Anat Embryol (Bevl) 191: 1–10
Wride MA, Lapchak PH, Sanders EJ (1994) Distribution of TNFα-like proteins correlates with some regions of programmed cell death in the chick embryo. Int J Dev Biol 38: 673–682
Wunder E, Fleischer-Reischmann B (1983) Response of lymphocytes from Fanconi's anemia patients and their heterozygous relatives to 8-methoxy-psoralene in a cloning survival test system. Hum Genet 64: 167–172
Wunder E, Mortensen BT, Schilling F, Henon PR (1993) Anomalous plasma concentrations and impaired secretion of growth factors in Fanconi's anemia. Stem Cells (Dayt) 11 [Suppl 2]: 144–149
Yamashita T, Barber DL, Zhu Y, Wu N, D'Andrea AD (1994) The Fanconi anemia polypeptide FACC is localized to the cytoplasm. Proc Natl Acad Sci USA 91: 6712–6716
Yonish-Rouach E, Resnitzky D, Lotem J, Sachs L, Kimchi A, Oren M (1991) Wild-type p53 induces apoptosis of myeloid leukaemic cells that is inhibited by interleukin-6. Nature 352: 345–247
Yoshimitsu K, Kobayashi Y, Usui T (1984) Decreased superoxide dismutase activity of erythrocytes and leukocytes in Fanconi's anemia. Acta Haematol 72: 208–210

Young BR, Painter RB (1989) Radioresistant DNA synthesis and human genetic diseases. Hum Genet 82: 113–117

Youssoufian H (1994) Localization of Fanconi anemia C protein to the cytoplasm of mammalian cells. Proc Natl Acad Sci USA 91: 7975–7979

Youssoufian H, Auerbach AD, Verlander PC, Steimle V, Mach B (1995) Identification of cytosolic proteins that bind to the Fanconi anemia complementation group C polypeptide in vitro. Evidence for a multimeric complex. J Biol Chem 270: 9876–9882

Yumoto Y, Hanafusa T, Hada H, Morita T, Oguchi S, Shinji N, Mitani T, Hamaya K, Koide N, Tsuji T (1995) Loss of heterozygosity and analysis of mutation of p53 in hepatocellular carcinoma. J Gastroenterol Hepatol 10: 179–185

Zambeti-Bosseler F, Scott D (1980) Cell death, chromosome damage and mitotic delay in normal human, ataxia telangiectasia and retinoblastoma fibroblasts after X-irradiation. Int J Radiat Biol 39: 547–558

Zhang Y, Harada A, Bluethmann H, Wang JB, Nakao S, Matsuchima K (1995) Tumor necrosis factor (TNF) is a physiologic regulator of hematopoietic progenitor cells: increase of early hematopoietic progenitor cells in TNF receptor p55-deficient mice in vivo and potent inhibition of progenitor cell proliferation by TNF alpha in vitro. Blood 86: 2930–2937

Zimmerman RJ, Chan A, Leadon SA (1989) Oxidative damage in murine tumor cells treated in vitro by recombinant human tumor necrosis factor. Cancer Res 49: 1644–1648

Genetic Alterations in Human Tumors

K.R. Cho and L. Hedrick

1 Introduction

At the level of the cell, cancer is essentially a genetic disease. Studies from multiple disciplines, including molecular biology, somatic cell genetics, genetic epidemiology, tumor virology, chemical carcinogenesis, and others, have converged over the last few decades to provide many fundamental insights into the genetic basis of cancer. These studies have shown that the genesis of malignant tumors is dependent on the accumulation of a series of DNA alterations that generally occur somatically, but can also be inherited in the germ line. These alterations have a cumulative effect on many important cellular processes that become dysfunctional in cancers, including growth control, differentiation, and interactions of the affected cell with its microenvironment.

Departments of Pathology, Gynecology and Obstetrics, and Oncology, The Johns Hopkins University School of Medicine, Room 659 Ross Research Building, 720 Rutland Avenue, Baltimore, MD 21205, USA

The genetic alterations responsible for tumor development and progression must, by some mechanism, confer a growth advantage to the cells in which they accumulate. To date, alterations have been shown to occur in three specific classes of cellular genes: oncogenes, tumor suppressor genes, and genes involved in DNA repair. The products of oncogenes function in a positive way to promote tumorigenesis. Their normal cellular counterparts, called proto-oncogenes, encode proteins that are important regulators of many aspects of cell growth. Thus, the activating mutations that generate oncogenic alleles are effectively "gain-of-function" mutations. In contrast, mutant tumor suppressor genes lose their ability to regulate growth; that is, mutations in these genes generally result in "loss of function." The proteins encoded by both proto-oncogenes and tumor suppressor genes have a myriad of functions, in part reflected by their varied locations in the cell, including the nucleus, cytoplasm, inner cell membrane, and cell surface. More recently, mutations in a third class of genes have been found to occur in human tumors. These genes encode proteins involved in the recognition and repair of damaged DNA. As with tumor suppressor genes, loss-of-function mutations appear to underlie the role of DNA repair genes in tumorigenesis. However, such mutations do not appear to directly affect growth control. Rather, their inactivation presumably causes an increased rate of mutation in oncogenes and tumor suppressor genes in affected cells.

The identification and characterization of specific genetic abnormalities in human tumors have confirmed many well-established hypotheses of tumorigenesis, revealed many new facts about tumor cells, and have led the way to a more thorough understanding of the pathogenesis of human cancer. Moreover, our knowledge of specific genetic changes in cancer cells is already being put to practical use with respect to the diagnosis, assessment of prognosis, and treatment of cancer patients.

It is not our purpose to provide a comprehensive review of all of the various mutations that have been described to date in human cancers. The field of cancer genetics is moving so quickly that such a categorization would certainly be out of date in a short period of time. Instead, we hope to accomplish the following: (a) to review the mechanisms by which proto-oncogenes are frequently activated in human tumors, each with illustrative examples – these mechanisms include point mutation, gene amplification, chromosomal translocation, and rearrangement; (b) to review the mechanisms by which tumor suppressor genes are frequently inactivated in human tumors, including point mutation with loss of the remaining wild-type allele (often detected as losses of heterozygosity, LOH); (c) to review some of the recent studies of inherited cancer syndromes, with emphasis on mutations in genes involved in the recognition and repair of damaged DNA; and (d) to briefly discuss some genetic alterations in cancer cells that do not necessarily target tumor suppressor genes or oncogenes, such as aneuploidy and DNA methylation. Our emphasis will be less on the molecular mechanisms through which these alterations arise than on the manner in which our understanding of these genetic alterations is impacting on the diagnosis and management of cancer patients today.

2 Genetic Alterations Frequently Affecting Oncogenes

Much of our knowledge of oncogenic alleles in human cancer is derived from studies of the transforming genes of RNA and DNA tumor viruses. Despite the fact that numerous proto-oncogenes were identified by such studies, only a relatively small subset of these cellular genes has been found to be frequently targeted by somatic mutations in human cancers. Although the great majority of mutations that give rise to oncogenes occur somatically, germ line transmission of dominant transforming genes (specifically, mutated *RET* genes) has recently been described in patients affected by the cancer syndromes multiple endocrine neoplasia (MEN)2A and (MEN)2B (SANTORO et al. 1995).

Several approaches have been used to identify human oncogenes. For example, chronic transforming retroviruses can alter the structure and/or expression of proto-oncogenes by integrating into or near these genes. Proto-oncogenes activated by this type of insertional mutagenesis include c-*MYC*, c-*ERBB*, c-*MYB*, c-*MOS*, *INT*1, and *INT*2, among others (reviewed in PERKINS and VANDE WOUDE 1993). Other oncogenes, such as H-*RAS*, were identified by transfer of tumor-derived DNA into non-tumorigenic cells. Yet additional oncogenes were identified by the characterization of non-random chromosomal abnormalities such as translocation breakpoints, extrachromosomal elements known as double-minute chromosomes, and novel banding regions identified in cytogenetic analyses. Tumor cell DNAs clearly contain many mutated genes, some that directly contribute to tumorigenesis and others that are altered as innocent bystanders in genetically unstable tumor cell. How might these "bystanders" be differentiated from genes causally involved in tumor development? Simply stated, the identity of a gene as an oncogene can be tested by introducing the candidate gene into normal cells and observing the cellular response. If the cell acquires at least some attributes of a cancer cell, then the candidate gene may be considered an oncogene (BISHOP 1985). Although there are 50 000–100 000 genes in the human genome, only about 50 genes identified thus far fit this functional definition. Of these 50, only 20 or so have been found to be mutated in human tumors (WEINBERG 1994). Specific examples of oncogenes activated in human tumors are described below.

2.1 *RAS* Genes Are Activated by Point Mutation

DNA transfer techniques were used to identify the first oncogenic *RAS* allele from a human bladder carcinoma cell line (KRONTIRIS and COOPER 1981; PERUCHO et al. 1981; SHIH and WEINBERG 1982; DER et al. 1982; PARADA et al. 1982). In these experiments (Fig. 1), DNA from the human bladder carcinoma cell line EJ (also known as T24) was introduced into immortalized but contact inhibited rodent fibroblasts. Those cells receiving transforming sequences from the human tumor DNA underwent morphological changes and loss of contact inhibition ("focus formation"). Recovery and characterization of the human tumor-derived sequences

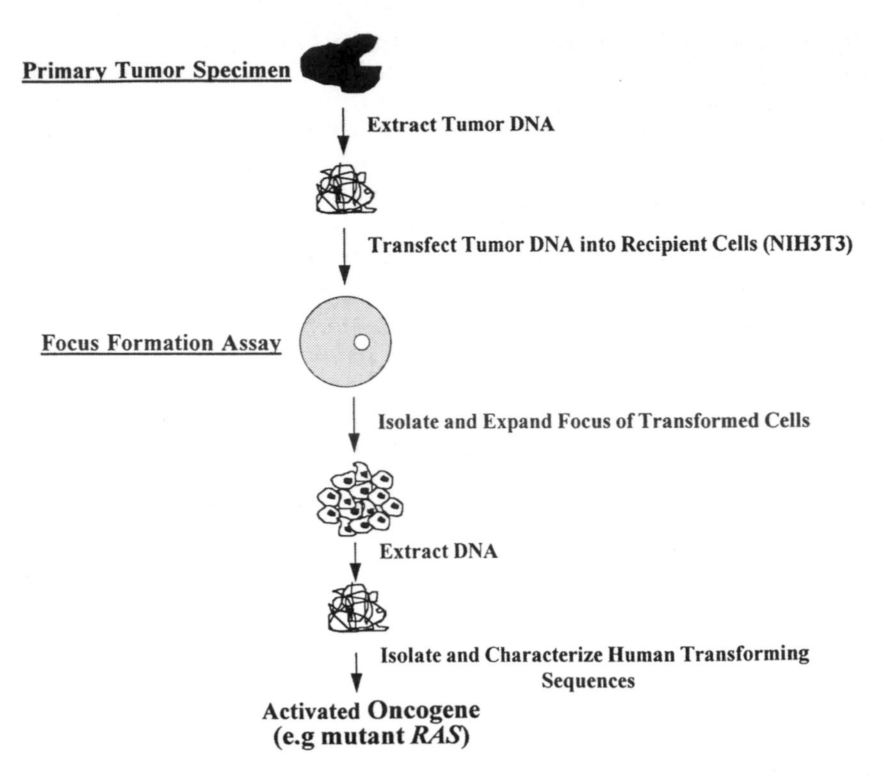

Fig. 1. DNA transfection assay for transforming sequences (oncogenes). DNA is isolated from a primary human tumor specimen or from cultured tumor cells. The DNA is then transfected into immortalized rodent cells that exhibit contact inhibition in culture (e.g., NIH3T3 cells). If the tumor DNA contains oncogenic alleles, the growth of cell(s) taking up the transforming sequences may fail to be inhibited by cell-cell contact, resulting in a "focus" of highly refractile cells that can be isolated and further expanded into a large population of transformed cells. The specific human tumor-derived DNA sequences responsible for transformation of the rodent cells can be identified by repeating this process with DNA isolated from a focus. In each round of such serial transfection and focus formation assays, the human transforming sequences are enriched with each round. This DNA transformation assay has successfully identified several oncogenes, most notably oncogenic *RAS* alleles

mediating this transformation resulted in the discovery of the first activated oncogene in human cancer, H-*RAS*. Further analysis of the activated *RAS* sequences established that the H-*RAS* allele from the bladder cancer cell line differed from the normal H-*RAS* gene by a single nucleotide substitution (point mutation) at codon 12 (Capon et al. 1983). Numerous additional studies identified point mutations in one of the three highly related *RAS* genes (H-*RAS*, K-*RAS*, or N-*RAS*) in approximately 10%–15% of human tumors (reviewed in Barbacid 1987). *RAS* gene mutations are even more common in certain cancer types, such as adenocarcinomas of the colon, pancreas, thyroid, and lung. The mutational spectrum and functional significance of *RAS* gene mutations have been reviewed in detail (Bos 1989). Overall, mutations at codon 12 of the K-*RAS* gene are most common, although mutations at codons 13 and 61 also occur. There is some evidence that environmental agents may contribute to the induction of the point mutations. For example, the majority of *RAS* gene mutations are G to A transitions at the second G of a GG pair (Bos 1989). This type of mutation is characteristic of those caused by alkylating agents.

What is the functional significance of these point mutations? Each of the *RAS* genes encodes a 21-kD guanine nucleotide-binding protein (p21). The ras proteins function in the transduction of growth and differentiation signals from activated transmembrane receptors to downstream protein kinases (reviewed in BOGUSKI and McCORMICK 1993; LOWY and WILLUMSEN 1993; McCORMICK 1993; BOURNE et al. 1990). Clearly, single nucleotide substitutions can have any of several consequences for the proteins encoded by genes altered by this type of mutation. Some mutations are missense; that is, they result in the substitution of one amino acid for another. Other point mutations, leading to the substitution of a stop codon for one encoding an amino acid and thereby creating a truncated protein, are called nonsense mutations. Point mutations can also occur in splice site recognition sequences, leading to the incorrect splicing of the RNA transcript (e.g., "exon skipping"). Such splice junction mutations can result in a frame shift that leads to truncation of the protein or may delete the portion of the protein encoded by the "skipped" exon. The oncogenic mutations at codons 12, 13, and 61 of the *RAS* genes are missense mutations, as might be expected for mutations that result in a gain of function. The RAS proteins encoded by these mutant genes have reduced ability to hydrolyze guanosine triphosphate (GTP) to guanosine diphosphate (GDP), primarily via their decreased ability to interact with the GTPase-activating protein called RAS-GAP. Hence, the mutant ras protein remains in the GTP-bound or activated state, no longer switching between the active (GTP-bound) and the inactive (GDP-bound) form. This scenario underscores the concept that proto-oncogenes do not lie dormant in the cell with the sole purpose of promoting tumorigenesis when activated. Rather, proto-oncogenes encode proteins that have important regulatory properties in normal cells and the mutant (oncogenic) proteins have altered function.

RAS gene mutations have been studied quite extensively in both pre-malignant (adenomas) and malignant colorectal tumors (VOGELSTEIN et al. 1988). K-*RAS* or N-*RAS* mutations are present in approximately 50% of adenomas greater than 1 cm and in a similar percentage of carcinomas. In contrast, *RAS* mutations are relatively uncommon in small (<1 cm) adenomas. These findings suggest that *RAS* gene mutations may be, at least in part, responsible for the progression of small adenomas to larger, more clinically aggressive lesions. However, it should be emphasized that *RAS* mutations are not always associated with malignant potential, as they have been identified in hyperplastic polyps and aberrant crypt foci (JEN et al. 1994; PRETLOW et al. 1993), lesions that have not clearly been shown to progress to colorectal cancer.

Our knowledge of *RAS* gene mutations has already been applied to the assessment of prognosis for patients with certain types of cancers. For example, K-*RAS* mutations are present in about 30% of adenocarcinomas in the lung (Bos 1989) and patients with tumors harboring mutant K-*RAS* genes have been shown to have a poorer prognosis than those with tumors containing wild-type K-*RAS* genes (SLEBOS et al. 1990). *RAS* mutation does not appear to be a useful prognostic marker in colorectal or other carcinomas in which *RAS* mutations are frequently present. Preliminary studies also suggest that the detection of *RAS* gene mutations may prove to be useful as an adjunctive tool in the early diagnosis of pancreatic and colorectal carcinomas since K-*RAS* gene mutations have been detected in the stool of patients with these tumor types (SIDRANSKY et al. 1992; CALDAS et al. 1994) as well as in fine-needle aspiration specimens of pancreatic cancers (URBAN et al. 1993; SHIBATA et al. 1990). Given the prevalence of *RAS* gene mutation in specific types of cancer as well as cancers in general, there is great interest in further studying *RAS*-dependent growth regulatory pathways. In particular, the identification of agents that inhibit the altered signaling of a mutant ras protein may ultimately lead to new treatment strategies for cancer patients.

2.2 *NEU* and *MYC* Genes Are Frequently Amplified in Certain Human Tumors

Cytogenetic studies of cancer cells have provided many of our insights into the genetic abnormalities associated with human cancers. In fact, several oncogenes were identified because they are contained within DNA sequences amplified in tumor cells. Such amplified DNA sequences are detected in karyotypic analyses as double-minute chromosomes (DMs) and novel banding regions known as homogeneously staining regions (HSRs). These DMs and HSRs typically contain 20 to several hundred copies of a specific chromosomal region. Selection for and maintenance of these amplified DNA sequences is thought to be driven by the increased copy number and concomitant increased expression of target gene(s) within the larger amplicon. Because the amplified region of DNA usually involves several hundred thousand base pairs, in some tumors more than one oncogene may be contained within the amplified segment (KHATIB et al. 1993). The mechanisms of gene amplification are discussed in detail by Tlsty (this volume).

The c-*MYC* proto-oncogene is the cellular homologue of a gene present in several highly oncogenic avian retroviruses. The c-*MYC* gene and its closely related cellular homologues N-*MYC* and L-*MYC* are frequently amplified in certain tumor types. For example, about 30%–40% of small cell carcinomas of the lung have amplification of c-*MYC*, N-*MYC* or L-*MYC* (DEPINHO et al. 1991). *MYC* gene amplification is also observed, albeit less frequently, in some other types of epithelial cancers, including carcinomas of the breast, colorectum, and cervix (DEPINHO et al. 1991; LITTLE et al. 1983; OCADIZ et al. 1987; BAKER et al. 1988). Amplification of *MYC* genes has also been identified in non-epithelial tumor types. Examples include c-*MYC* amplification in promyelocytic leukemia (COLLINS and GROUDINE 1982; DALLA-FAVERA et al. 1982) and N-*MYC* amplification in a subset of glioblastomas and neuroblastomas (BRODEUR et al. 1984; FULLER and BIGNER 1992; RASHEED and BIGNER 1991). N-*MYC* gene amplification is observed more often in patients with high-stage neuroblastomas, and patients with neuroblastomas that show N-*MYC* amplification have a much poorer prognosis than patients whose tumors lack this genetic alteration (Brodeur et al. 1984; Schwab and Amler 1990). Similarly, amplification of *MYC* genes in small cell lung carcinoma also seems to be associated with disease progression (Garte 1993). The correlation between *MYC* amplification and prognosis of patients with neuroblastomas and small cell carcinomas of the lung demonstrates additional clinical settings in which identification of specific genetic abnormalities impacts on the management of cancer patients.

How do the *MYC* genes normally participate in the regulation of cell growth? Recent studies have shown that each of the myc proteins dimerizes with a partner protein called max (Blackwood and Eisenman 1991). The resultant heterodimers then function as transcription factors by binding specific DNA sequences. The three *MYC* genes, although closely related, each have a unique pattern of expression during development and show subtle functional differences in in vitro assays. Moreover, amplification of *MYC* genes in certain tumor types does not appear to be interchangeable; for example, neuroblastomas show frequent amplification of N-*MYC*, but not of c-*MYC* or L-*MYC*. Taken together, these findings suggest that each *MYC* gene product functions to regulate expression of its own set of downstream genes, which are presumably involved in cell proliferation. Such specific target genes have yet to be identified.

The *NEU* gene (also known as *HER*-2 or c-*ERB*B-2) is another proto-oncogene that is frequently activated in human tumors through gene amplification and concomitant overexpression of its mRNA and protein. *NEU* amplification occurs most frequently in carcinomas of the breast, ovary, and stomach (Slamon et al. 1989; Zhou et al. 1988; Paik 1992). The presence or absence of *NEU* amplification has also been found to be of prognostic significance in patients with node-negative breast carcinomas. Studies have shown that patients whose tumors showed *NEU* amplification and/or overexpression had a poorer clinical outcome than patients whose tumors did not demonstrate this genetic change (Slamon et al. 1989; Press et al. 1993).

The *NEU* gene encodes a cell surface protein quite similar to the epidermal growth factor receptor (EGFR). The neu protein has a single membrane-spanning domain and a tyrosine kinase functional domain in the cytoplasmic portion. Like the *EGFR* gene, the *NEU* gene is believed to encode a peptide hormone receptor. In fact, it appears that EGFR and neu can form heterodimers, allowing neu activity to be modulated by EGFR agonists such as epidermal growth factor (EGF) and transforming growth factor-α (TGF-α) (Dougall et al. 1994). Numerous studies are underway to identify specific neu ligands, to understand the role of neu in growth regulation, and to elucidate the role of *NEU* gene overexpression in the pathogenesis of certain cancers. Ultimately, identification of *NEU* gene amplification and overexpression may find clinical utility not only in the assessment of prognosis of cancer patients, but perhaps in their treatment. In experimental model systems, antibodies directed against the extracellular domain of neu have been shown to inhibit tumor cell growth both in vitro and in animal tumor models (Dougall et al. 1994; Drebin et al. 1988).

2.3 Other Oncogenes, Such as c-*ABL* and *BCL*-2, Are Activated by Chromosomal Translocations

Cytogenetic studies have revealed several recurring and highly consistent chromosomal translocations in certain tumor types. With the improvement of cytogenetic methods and the introduction of molecular biological tools, several oncogenes at or spanning the chromosomal breakpoints have been identified (reviewed in RABBITTS 1994). Cloning and characterization of these breakpoints have revealed some general principles underlying the functional consequences of alterations of the affected genes. There appear to be two major results of chromosomal translocations: (a) the gene encoding a T cell receptor (TCR) or an immunoglobulin (Ig) comes to lie near a proto-oncogene, resulting in overexpression of the proto-oncogene's mRNA and protein; or (b) the chromosomal breaks occur within two specific genes, resulting in a fusion gene encoding a chimeric protein with altered function. The fused genes often encode transcription factors such that the translocations result in altered transcription of downstream target genes in the affected tumor cells. Translocations affecting TCR or Ig genes and rearrangements creating fusion genes are frequently identified in hematopoietic malignancies. In solid tumors, the most common consequence of chromosomal translocation is the formation of chimeric proteins.

The *BCL*-2 (B cell lymphoma-2) gene was identified because it was found to be activated by a translocation between the Ig heavy chain locus on chromosome 14q and sequences on chromosome 18q. This translocation juxtaposes the *BCL*-2 gene on chromosome 18q near the regulatory elements at the Ig locus (TSUJIMOTO et al. 1985; CLEARY et al. 1986), causing deregulation of *BCL*-2 gene expression and overproduction of *BCL*-2 mRNA and protein. This type of translocation is typically noted in a subset of non-Hodgkin's lymphomas, namely the follicular lymphomas, which are of B cell origin. Detection of this translocation (traditionally by Southern blot analysis) has been used as an adjunctive tool in the diagnosis of these malignancies, particularly in complicated cases. Until relatively recently, the classification of hematopoietic malignancies was based exclusively on the morphology of the neoplastic cells. As might be expected, proper classification of leukemias and lymphomas was largely dependent on the skill of the pathologist and was limited by problems with reproducibility. Molecular DNA analyses detecting chromosomal translocations typical of each tumor type have contributed greatly to our understanding and appropriate classification of lymphoproliferative disorders in general. More recently, polymerase chain reaction (PCR)-based methods have been developed to assist in establishing the primary diagnosis and to detect minimal residual disease in patients with several types of hematopoietic neoplasms (reviewed in BORDEN et al. 1993; POTTER et al. 1993; YUNIS and TANZER 1993; GRIESSER 1993).

Simply stated, cancers are excessive accumulations of tumor cells which presumably result from an imbalance in cell proliferation and cell death, or some combination of both. Unlike many other oncogenes, which function to increase the percentage of cells in the population that are proliferating, *BCL*-2 appears to contribute to tumorigenesis by protecting tumor cells from apoptosis (programmed cell death). Early clues supporting this notion were provided by studies showing that survival of interleukin-3-dependent lymphoid cells is prolonged in the setting of constitutive expression of *BCL*-2, even though there is no concomitant increase in the number of cells proliferating (VAUX et al. 1988). In addition, T cells expressing a *BCL*-2 transgene display sustained viability and resistance to lymphotoxic agents (SENTMAN et al. 1991; STRASSER et al. 1991). The mechanisms through which *BCL*-2 mediates the suppression of apoptosis remain incompletely defined; however, it appears that *BCL*-2 may regulate antioxidant pathways and cause a decrease in the production of reactive oxygen species (HOCKENBERY et al. 1993; KANE et al. 1993). The function of bcl-2 protein may be modulated by its interaction with other cellular proteins such as bcl-x, bax, and r-ras (reviewed in REED 1994).

Although some proto-oncogenes are activated by juxtaposing them near genes that normally undergo rearrangement, many others are activated by translocations resulting in gene fusion, which results in the expression of a chimeric protein of altered function. An example of such a gene fusion is provided by the translocation between chromosomes 9 and 22 observed in the great majority of chronic myelocytic leukemias. In this translocation (known as the Philadelphia chromosome) the proto-oncogene c-*ABL* on chromosome 9q is translocated to a 5-kb region on chromosome 22q called the breakpoint cluster region (*BCR*) which contains a transcription unit (HEISTERKAMP et al. 1983; BARTRAM et al. 1983; DE KLEIN et al. 1982). The consequence of this translocation is fusion of the promoter and 5' sequences of the *BCR* locus with c-*ABL*. The 8.2-kb fusion mRNA predicts a novel fusion protein of 210 kDa (COLLINS et al. 1984; KONOPKA et al. 1984). Assays detecting the bcr-abl fusion protein have been used as adjunctive tools in the diagnosis of chronic myelogenous leukemia. A second type of Philadelphia chromosome is observed in a small percentage of patients with chronic myelogenous leukemia and a subset of patients with acute lymphoblastic leukemia (DE KLEIN et al. 1986). In this translocation, the break on chromosome 22 falls within a *BCR* intron that is outside the main breakpoint cluster region. The resultant *BCR-ABL* fusion mRNA is 7.0 kb and encodes a 185-kD chimeric protein. Patients whose leukemic cells exhibit this variant Philadelphia chromosome have been shown to have a worse prognosis than those with the typical 9–22 translocation (PERKINS and VANDE WOUDE 1993).

The mechanisms through which the bcr-abl fusion protein contributes to leukemogenesis are not clearly understood. The normal c-*ABL* gene product is a member of the tyrosine kinase family and presumably participates in signal transduction. It appears that *BCR*-encoded sequences in the chimeric protein function to activate the kinase activity, perhaps through direct binding of the *ABL* kinase regulatory domain to an SH2-binding region in *BCR* (PENDERGAST et al. 1991; LUGO et al. 1990).

3 Genetic Alterations Frequently Affecting Tumor Suppressor Genes

Several tumor suppressor genes have been discovered in the last few years. Although study of these genes lags behind oncogene research by at least a decade, it is already clear that their inactivation contributes substantially to cancer causation. In fact, cancers seem to arise when there is disruption of the delicate balance between the actions of the proteins encoded by the growth-promoting proto-oncogenes and the growth-controlling tumor suppressor proteins. The cell fusion studies by HARRIS and colleagues provided the initial basis for the concept that the ability to generate a tumor is essentially a recessive trait at the cellular level (reviewed in HARRIS 1988). These investigators found that fusion of malignant murine tumor cells with non-tumorigenic murine cells often resulted in hybrids that were non-tumorigenic (Fig. 2). When passaged extensively in culture, the suppressed hybrids often reverted to tumorigenicity, and these tumorigenic revertants were found to have lost chromosomes when compared to their non-tumorigenic counterparts. Studies of hybrids generated by fusion of human tumor cells with normal cells yielded similar results. The observation that chromosome losses were asso-

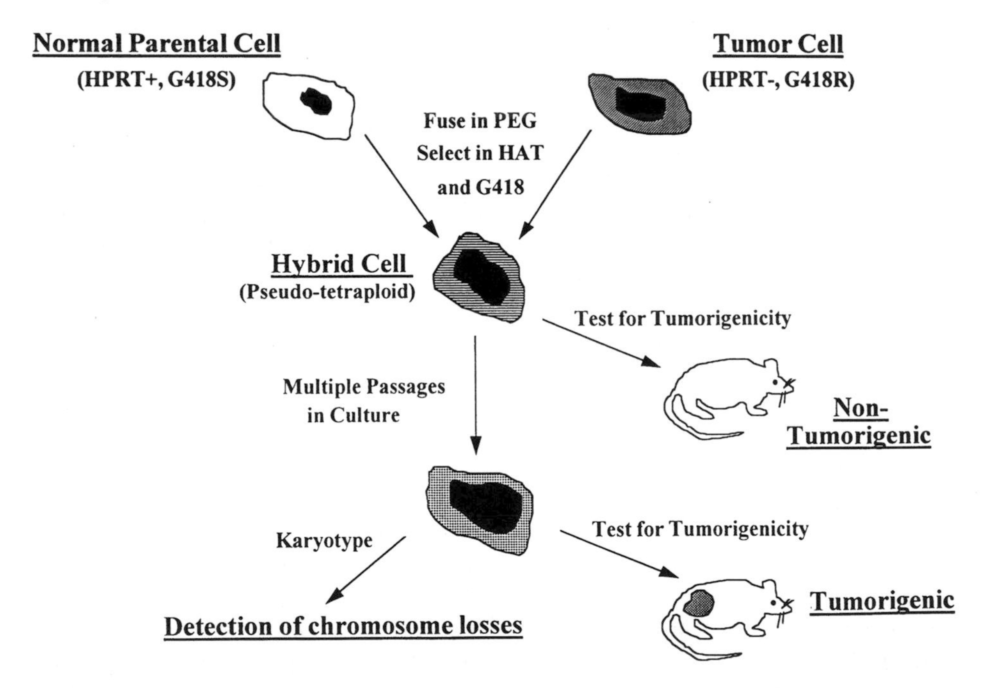

Fig. 2. Somatic cell genetic evidence for tumor suppressor genes. Somatic cell hybrids can be generated by fusing malignant cells with non-tumorigenic cells under the appropriate conditions. Hybrids are identified by growth in media that selects for cells containing the genetic complement of both parental cells. These pseudotetraploid cells are often non-tumorigenic when injected into mice. However, when passaged extensively in culture, the hybrid cells often revert to tumorigenicity. These tumorigenic revertants show losses of specific chromosomes when compared to the non-tumorigenic hybrids

ciated with reversion to tumorigenicity suggested the hypothesis that specific chromosomes, or even single genes on specific chromosomes, might be responsible for the tumor suppression. This hypothesis was confirmed by subsequent experiments showing that introduction of single specific chromosomes into certain tumorigenic cells resulted in suppression of tumorigenicity (WEISSMAN et al. 1987; OSHIMURA et al. 1990; TRENT et al. 1990). However, some specificity regarding the ability of a particular chromosome to suppress tumorigenicity of specific tumor types was noted. For example, tumor suppression was observed following introduction of chromosome 11 into cervical carcinoma cells or rhabdomyosarcoma cells, while chromosome 11 failed to suppress the tumorigenicity of some neuroblastomas and renal cell carcinomas (OSHIMURA et al. 1990; SHIMIZU et al. 1990).

While these cell fusion experiments suggested a critical role for the inactivation of cellular genes in tumor formation, concurrent epidemiological and genetic studies provided additional evidence supporting the existence of tumor suppressor genes. Amongst these, the studies of KNUDSON (1971, 1985) are of particular importance since they laid the groundwork for many of the concepts and techniques used to identify and characterize tumor suppressor genes and the mechanisms by which they are inactivated in both germ line and somatic cells.

Retinoblastomas are relatively uncommon malignant eye tumors affecting children. They occur sporadically in many patients, but are also observed in some families in a pattern consistent with autosomal dominant inheritance. KNUDSON analysed the age-specific incidence of retinoblastoma and proposed that two mutagenic events were required for the development of retinoblastoma, both in sporadic and inherited cases (the so-called two-hit hypothesis). In patients with the inherited form of the disease, one mutation was inherited and therefore present in all cells of the affected individual, including developing retinoblasts. A second mutation occurring in any of these retinoblasts would then lead to the development of a retinoblastoma. In those patients with the sporadic form of the tumor, both mutations must be acquired somatically and must occur in the same developing retinoblast. The low probability that two separate somatic mutations would occur in the same retinoblast is consistent with the low prevalence of retinoblastoma in the general population. In contrast, patients with the inherited form of the disease typically develop one or more retinoblastomas depending on the frequency of the second mutational event(s). KNUDSON subsequently proposed that the two mutations were likely to inactivate the same gene rather than two different ones. Thus, although predisposition to retinoblastomas is inherited by the affected individual in a dominant fashion, mutations in the putative disease gene are predicted to be recessive at the cellular level, such that both copies of the same gene need to be inactivated for a tumor to develop. Cytogenetic studies provided the first clues to the localization of the retinoblastoma predisposition gene to chromosome 13 (FRANCKE 1976). Interstitial deletions of band q14 of chromosome 13 were observed in the normal (non-neoplastic) cells of a subset of patients affected by the inherited form of retinoblastoma. Other studies provided evidence for strong genetic linkage between the esterase D locus at chromosome 13q14 and the retinoblastoma predisposition locus (SPARKES et al. 1983).

3.1 Tumor Suppressor Genes Are Often Inactivated by Loss of Heterozygosity

Somatic cells contain two copies of each autosomal chromosome, one inherited from each parent. Probes detecting DNA sequence polymorphisms allow investigators to distinguish between the maternally and paternally derived copies of particular DNA sequences in both normal and neoplastic cells. Such molecular biological tools were subsequently used to identify frequent losses of one parental copy of chromosome 13 in retinoblastomas (FUNG et al. 1987) regardless of whether they were inherited or sporadic. These losses, termed losses of heterozygosity (LOH), were detected by analysis of chromosome 13 DNA polymorphisms in matched pairs of tumor and normal DNA. LOH, also known as allelic loss, was found to occur through a number of different mechanisms (Fig. 3) including (a) chromosome non-disjunction with reduplication of the remaining chromosome; (b) chromosome non-disjunction without reduplication of the retained chromosome; and (c) mitotic recombination. In the inherited tumors showing chromosome 13 LOH, it was determined that the retained copy of chromosome 13 was always derived from the affected parent, suggesting that the LOH was effectively "unmasking" a pre-existing recessive mutation at the putative tumor suppressor gene locus. Taken together, these studies served to support KNUDSON's two-hit hypothesis and demonstrated that similar genetic mechanisms could account for both inherited and sporadic cases of certain cancers.

Cloning of the retinoblastoma susceptibility gene (*RB*1) was facilitated by the identification of a chromosome 13 marker that detected rearrangement and/or homozygous loss (loss of both copies) of chromosome 13q14 sequences in Southern blot analyses of some primary retinoblastomas (DRYJA et al. 1986). These findings ultimately led to the identification of the *RB*1 gene, a large gene containing 27 exons and spanning over 200 kb of genomic DNA (FUNG et al. 1987; LEE et al. 1987; FRIEND et al. 1986). Once cloned, investigators were able to more definitively study both inherited and somatic *RB*1 mutations. Although gross deletions and rearrangements of the *RB*1 gene are observed in a subset of retinoblastomas, most tumors lack gross alterations of the gene (DUNN et al. 1989; YANDELL et al. 1989; GODDARD et al. 1988). Rather, the majority of mutations inactivating the *RB*1 gene are more subtle, including point mutations, small insertions, and deletions. The effects of such mutations are discussed below.

3.2 Tumor Suppressor Genes Are Often Inactivated by Point Mutations and Other Subtle DNA Alterations

As described in the section on oncogenes, point mutations in tumor suppressor genes can have any of several consequences for the proteins encoded by the mutant genes. The mutations affecting *RB*1 typically result in premature truncation of the rb protein. Premature truncation of proteins may be caused by any of several

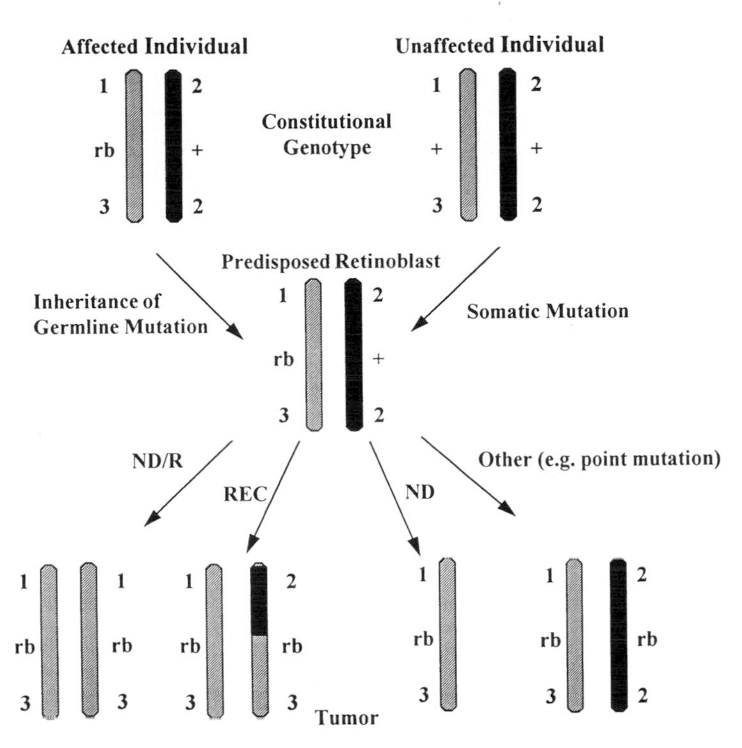

Fig. 3. Mechanisms that result in loss of heterozygosity for alleles at the retinoblastoma predisposition (*RB1*) locus. In the inherited form of the disease (*top left*), the affected patient has inherited one mutant *RB1* allele (*rb*) and one wild-type *RB1* allele (+). Thus, all of the cells in the affected patient, including the retinoblasts, contain one wild-type and one mutant allele such that the constitutional genotype at the *RB1* locus is (rb/+). In contrast, the constitutional genotype of the unaffected patient (*top right*) is +/+ (i.e., both alleles are wild-type). The two copies of chromosome 13 in non-neoplastic cells can be distinguished using polymorphic DNA markers flanking the *RB1* locus (these various polymorphic alleles are designated by *numerals*). In patients with the inherited form of the disease, a retinoblastoma arises from a predisposed retinoblast after inactivation of the remaining wild-type *RB1* allele. Patients with the sporadic form of retinoblastoma require somatic inactivation of both *RB1* alleles in the same developing retinoblast in order for a tumor to form. Several genetic mechanisms have been found to somatically inactivate wild-type *RB1* alleles during tumor development. These include: chromosome non-disjunction and reduplication of the remaining chromosome (*ND/R*), mitotic recombination (*REC*), non-disjunction (*ND*) without reduplication of the remaining chromosome, and other more localized mutations that inactivate the remaining *RB1* allele (*Other*)

different types of mutations in the corresponding DNA. For example, individual nucleotide substitutions may change a codon from one encoding an amino acid to one specifying translation termination (so-called nonsense mutations). Small insertions or deletions within the coding portion of the gene interrupt the open reading frame and often cause a frameshift, such that the DNA downstream of the insertion or deletion specifies incorrect amino acids followed by an early translation termination codon. Mutations affecting the splicing recognition signals at the intron-exon boundaries within a gene can also result in premature protein truncation. Such mutations cause the affected exons(s) to be skipped and other exons to be incorrectly spliced together. This aberrant splicing may also result in a frameshift, with the same consequences for the protein as outlined above. *RB*1 is not the only tumor suppressor gene known to be frequently inactivated by mutations that cause early protein truncation. The *APC* (adenomatous polyposis coli) tumor suppressor gene on chromosome 5q is also typically inactivated by such mutations (Nishisho et al. 1991; Groden et. al. 1991). Germ line mutations of *APC* have been identified in the great majority of patients affected by familial adenomatous polyposis or its variant, Gardner's syndrome. Moreover, 5q LOH and/or mutations of the *APC* gene are observed in the majority (60%–70%) of sporadic colorectal adenomas and carcinomas, suggesting that inactivation of *APC* is of critical importance in the development of epithelial colorectal tumors in general and not only in polyposis patients (Powell et al. 1992). It is not difficult to envision that prematurely truncated proteins are likely to be inactive, particularly if they are small or truncated early. Investigators have capitalized on the fact that *APC* is typically inactivated by nonsense mutations in developing assays that screen for truncated APC proteins in colorectal tumors and non-neoplastic cells from patients at risk for familial adenomatous polyposis (Powell et al. 1993). Such approaches are particularly advantageous since they obviate the need for laborious DNA-based sequence analyses of *APC*, which is a rather large gene.

Tumor suppressor genes may also be "inactivated" by missense point mutations, that is, mutations that cause substitution of one amino acid for another. The effects of missense mutations on tumor suppressor protein function are best illustrated by a brief review of the spectrum of mutations identified in the *p53* gene, the most commonly mutated gene known to date in human cancers. This gene lends itself well to mutation analysis since (a) it is relatively small (11 exons, 393 amino acids); (b) point mutations are distributed over a large region of the gene, allowing correlation between specific mutations and effects on protein function; and (c) the mutations are diverse (e.g., compared to those affecting the *RAS* genes), permitting more extensive inferences to be made regarding etiologic mechanisms. The *p53* gene encodes a 53-kD phosphoprotein that appears to function primarily in transcriptional regulation, although it may have other activities (Vogelstein and Kinzler 1992; Friend 1994). The p53 protein contains three major functional regions, an amino terminal acidic domain with transcriptional activation function, a central hydrophobic core domain that is a primary determinant of p53 conformation and sequence-specific DNA binding activity, and a carboxy terminal domain containing the basic nuclear localization sequence and sequences mediating homo-oligomerization of p53 protein. The five distinct regions of p53 sequence that are highly conserved are found within the DNA binding and transactivation domains. Not suprisingly, the *p53* gene mutations that have been characterized in human tumors cluster within these conserved regions, particularly in the portions of the gene encoding the DNA-binding domain. Thus, the vast majority of mutant *p53* alleles that have been studied inactivate the DNA-binding activity of the p53 protein.

The spectrum of *p53* gene mutations in human cancers has been extensively reviewed (Greenblatt et al. 1994). Missense mutations are most common in the conserved mid-region (79% of mutations) but are uncommon in the amino and carboxy termini (23% of mutations), where nonsense mutations predominate. This observation suggests that single amino acid substitutions in the amino and carboxy terminal domains do not typically alter p53 protein conformation or function to the same degree as similar changes in the hydrophobic mid-portion. In fact, the most commonly mutated amino acids in the DNA-binding domain are close to the protein-DNA interface as determined by the recently described crystal structure of the p53 protein (Cho et al. 1994).

Evaluation of the *p53* mutational spectrum has also provided some clues and confirmed previous suspicions regarding the etiology of certain cancer types. A particularly illuminating example is provided by analysis of *p53* gene mutations in basal cell and squamous carcinomas of the skin. Development of these tumors has been clearly linked to UV light exposure. UV radiation acts as a physical mutagen that produces distinctive pyrimidine dimers that result in characteristic tandem mutations (usually CC → TT transitions) if left unrepaired. These tandem mutations, while rare in other tumor types, are common in both cutaneous basal cell and squamous carcinomas – directly incriminating UV light as the cause of these mutations in skin cancers (Brash et al. 1991).

p53 mutation analysis has also found several clinical applications. For example, some studies have addressed the value of *p53* mutation as a prognostic factor in various tumor types. Expression of mutant p53 protein has been implicated as an independent prognostic indicator in carcinomas of the breast, stomach, colorectum, bladder, and lung (non-small cell types) (reviewed in Dowell and Hall 1994). Other studies have used *p53* mutation analysis to screen body fluids for recurrent cancers (e.g., urine for recurrent bladder cancer) (Sidransky et al. 1991), and to evaluate surgical margins of resection for small foci of tumor undetected by standard histopathological assessment (Brennan et al. 1995). Clearly, evaluation of *p53* mutations in human tumors has many diagnostic, prognostic, and therapeutic implications.

4 Mutations of DNA Mismatch Repair Genes and the Mutator Phenotype

Most adult human solid tumors, when properly analyzed, have been shown to have multiple mutations (Fearon and Vogelstein 1990). This finding strongly supports the concept that human tumors develop due to the acquisition and subsequent accumulation of multiple mutations. Loeb (1991) has cogently argued that the spontaneous mutation rate in human cells is not high enough to account for the number of mutations thought to be necessary for the development of neoplasia. Thus, he hypothesized that the establishment of a "mutator phenotype" may be a necessary and early step in the formation of human tumors. This mutator phenotype was defined as an increase in the mutation rate of affected cells, thus predisposing them to the accumulation of genetic alterations necessary in the pathogenesis of neoplasia (Loeb 1991). Several classes of genes were proposed that, when mutated, could generate the mutator phenotype. One of the suggested classes of genes included genes encoding DNA mismatch repair proteins.

Recently, the convergence of several different lines of scientific investigation have culminated in the identification and cloning of several human DNA mismatch repair genes (Fishel et al. 1993; Leach et al. 1993; Papadopoulos et al. 1994; Bronner et al. 1994). Moreover, it has been shown that these DNA mismatch repair genes are mutated in some human tumors and in the germ line of individuals affected by certain types of inherited predisposition to cancer. The story unfolded through an unforeseen combination of studies aimed at understanding human disease and fundamental processes in microorganisms and provides an extraordinary example of the potential impact of basic science on medical research.

Simultaneously, investigators studying both inherited and sporadic colorectal cancers noted alterations in the length of microsatellite sequences in DNA from tumor cells when compared to DNA isolated from the same patient's normal cells (Ionov et al. 1993; Thibodeau et al. 1993; Aaltonen et al. 1993; Peltomaki et al. 1993b) (Fig. 4). Several groups were working to understand the genetic basis of an inherited family cancer syndrome called hereditary non-polyposis colorectal cancer (HNPCC). HNPCC is an autosomal dominant disease with three or more family members affected in two successive generations, including at least one affected individual under the age of 50 years. The most common tumors found in affected individuals are colorectal and endometrial carcinomas. Experiments revealed that DNA isolated from virtually all of the tumors from affected HNPCC family members showed instability at multiple microsatellite loci (Aaltonen et al. 1993). The groups studying sporadic colorectal carcinomas simultaneously noted instability of microsatellite DNA sequences (Thibodeau et al. 1993; Ionov et al. 1993) in tumor DNA. Unlike tumors from individuals affected by HNPCC, in which nearly all of the analyzed tumors showed instability, only approximately 20% of the sporadic colorectal tumors studied demonstrated the phenotype. The phenomenon was referred to as microsatellite instability (MI) or replication error (RER). Initially, the significance of microsatellite instability and its potential role in tumorigenesis were unclear.

Shortly after the identification of microsatellite instability in human tumors, it was reported that mutations in DNA mismatch repair genes in *Saccharomyces cerevisiae* cause a 100–700-fold increase in the instability of simple dinucleotide repeat sequences in vivo (Strand et al. 1993). In addition, previous work in *Escherichia coli* and *S. cerevisiae* had shown that defects in the DNA mismatch repair system could result in instability of simple repetitive DNA sequences (reviewed in Modrich 1991). These observations provided a critical link between microsatellite instability and defects in DNA mismatch repair that set the stage for the identification of the human genes. Within several months a human homologue of the bacterial and yeast DNA mismatch repair genes was cloned using a PCR-based strategy that capitalized on DNA sequences conserved among the known microbial genes (Fishel et al. 1993; Leach et al. 1993). Shortly thereafter, three additional human homologues were cloned using a similar strategy, aided in part by the

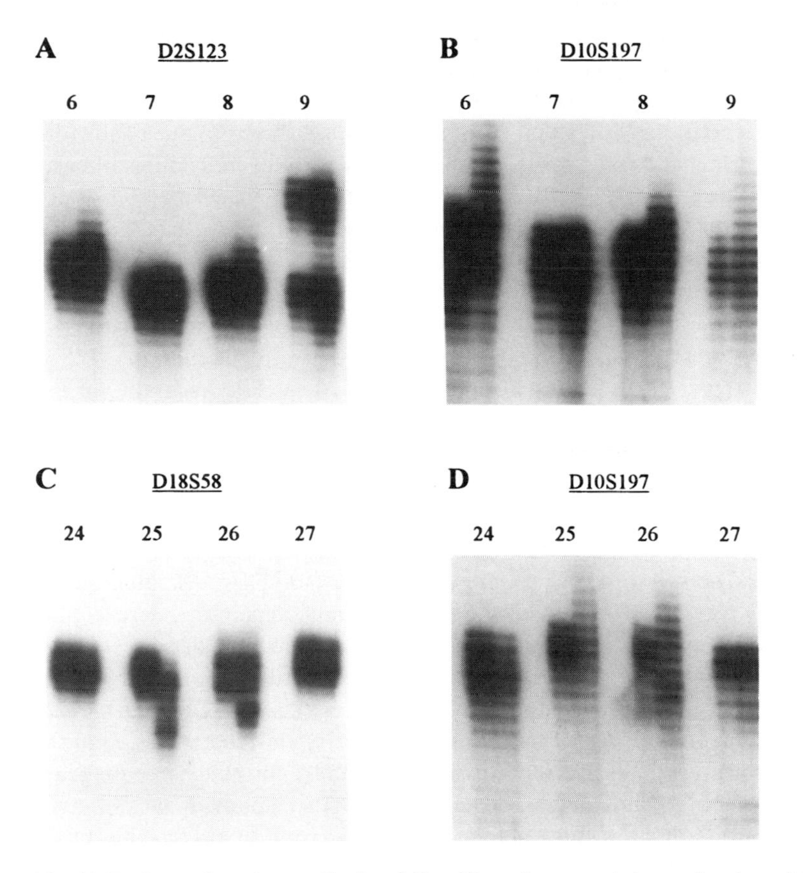

Fig. 4A–D. Assays for microsatellite instability. The polymerase chain reaction is used to amplify microsatellite sequences in tumor (T) vs. normal (N) DNA from the same patient. Tumors demonstrating microsatellite instability show alterations in the length of microsatellite sequences in the DNA from tumor cells when compared to matched normal DNA. The *panels* show several different normal-tumor DNA pairs assayed for microsatellite instability at different chromosomal loci (*D2S123*, *D10S197*, and *D18S58*). Cases 6, 8, 9, 25 and 26 exhibit the microsatellite instability phenotype

identification of conserved sequences in the data base of expressed sequence tags (dbEST) (Bronner et al. 1994; Papadopoulos et al. 1994).

The cloned human DNA mismatch repair genes have been named *hMSH2, hMLH1, hPMS1,* and *hPMS2,* in keeping with their microbial homologues. Each human DNA mismatch repair gene has been assigned a chromosomal location by physical mapping methods. The location of each gene has been shown to correlate with the location of the genes responsible for HNPCC as determined by genetic linkage analyses. The four genes, *hMSH2, hMLH1, hPMS1,* and *hPMS2,* are located on chromosomes 2p, 3p, 2q, and 7q, respectively. Studies of large families have determined that HNPCC is a genetically heterogeneous disease with some families demonstrating linkage to either chromosome 2p and 3p and other families showing linkage to the other chromosomal locations (Peltomaki et al. 1993a;

LINDBLOM et al. 1993; NYSTROMLAHTI et al. 1994). The combination of the linkage and physical mapping data provided further evidence for the role of DNA mismatch repair genes in the pathogenesis of neoplasms arising in the setting of HNPCC. It has since been shown that germ line mutations occur in one of the four known human mismatch repair genes in the great majority of individuals affected by HNPCC (FISHEL et al. 1993; LEACH et al. 1993; BRONNER et al. 1994; PAPADOPOULOS et al. 1994; NICOLAIDES et al. 1994). As expected, the specific gene mutated in each family is located on the respective chromosome to which the disease locus demonstrates genetic linkage.

Concurrent with the analyses of mutations in HNPCC kindreds, in vitro functional studies were performed that provided further evidence for the role of DNA mismatch repair genes in the production of microsatellite instability in human tumor cell lines. Using a microsatellite instability assay similar to that used in *S. cerevisiae,* investigators showed that the rate of instability of a dinucleotide repeat was a least two orders of magnitude higher in a colorectal carcinoma cell line derived from an MI-positive tumor compared to a cell line derived from a tumor that was MI negative (PARSONS et al. 1993). Such MI-positive colorectal carcinoma cell lines were also shown to be deficient in an early step in DNA mismatch repair, lending further credence to the role of DNA mismatch repair in microsatellite instability in human tumors. Moreover, a lymphoblastoid cell line derived from an affected HNPCC individual was able to appropriately direct mismatch repair in vitro, supporting the hypothesis that, at the cellular level, mismatch repair genes act in a recessive manner, requiring the inactivation of both alleles for expression of the phenotype. Accordingly, somatic mutations in the DNA mismatch repair genes have been identified in the tumors of affected HNPCC individuals. Affected members of HNPCC families thus inherit one mutant copy of the specific DNA mismatch repair gene responsible for the disease in their family and somatically acquire the mutation in the remaining normal allele. This situation is very similar to other familial cancer syndromes caused by inherited mutations of tumor suppressor genes, including familial adenomatous polyposis, and the inherited forms of Wilms' tumor and retinoblastoma. More recently, a subset of HNPCC patients were found to have widespread microsatellite instability not only in their tumor DNA, but also in the DNA of their non-neoplastic cells (PARSONS et al. 1995). This phenomenon occurred only in patients with certain germ line mutations of *hPMS2* and *hMLH1* not in patients with mutant *hMSH2* or *hPMS1* genes. These data suggest that some *hPMS2* and *hMLH1* mutations may confer dominant negative activity to the mutant proteins such that a mismatch repair defect is manifest when only one of the two copies of these genes is mutant.

How do microsatellite instability and associated mutations in DNA mismatch repairs genes predispose cells to the development of neoplasia? This question was addressed, in part, by looking for mutations in known cancer genes in MI-positive tumors from individuals with HNPCC. Many of the genetic alterations that underlie the development of colorectal carcinoma have been well characterized and are easily analyzed. Studies of MI-positive colorectal cancers showed that mutation frequencies of *p53*, K-*RAS*, and *APC* are similar to those found in sporadic forms of colorectal cancer, the majority of which are MI negative (AALTONEN et al. 1993). This finding supports the idea that tumors arising in HNPCC share similar pathogenetic mechanisms to colorectal tumors occurring in the sporadic form. Furthermore, it may explain why individuals who inherit a defective DNA mismatch repair gene so frequently develop tumors, as defects in the DNA mismatch repair system may lead to a marked increase in the rate at which mutations occur in genes that give rise to the neoplastic phenotype (e.g., tumor suppressor genes and oncogenes). Therefore, cells of individuals who inherit one mutant copy of a DNA mismatch repair gene and undergo somatic inactivation of the remaining allele will rapidly accumulate mutations in the critical genes that promote tumorigenesis.

The DNA mismatch repair system has been well characterized in microorganisms where it has been shown to detect and repair mispaired bases present in the cellular genome (MODRICH 1991). Thus, it is of obvious importance in controlling the fidelity of DNA replication by decreasing the number of mutations that become fixed in the genome following each round of DNA synthesis. When the DNA mismatch repair system is lacking, the rate at which mutations accumulate is increased (COX 1976). Although the DNA mismatch repair system has been much less well characterized in mammalian cells, there is substantial data supporting the hypothesis that its functions are very similar to those described in microorganisms. In addition to the genetic and molecular studies described above, biochemical studies have demonstrated that purified hMSH2 protein binds to mismatched microsatellite DNA sequences (FISHEL et al. 1994). Furthermore, as microsatellite DNA sequences are quite prone to errors during DNA replication, it follows that these sequences would demonstrate numerous alterations in the absence of an intact DNA mismatch repair system. These studies suggest that, in mammalian cells, microsatellite instability may simply serve as a marker of the mutator phenotype, and the neoplastic process is produced by an increased rate of mutations in oncogenes and tumor suppressor genes.

Since the initial finding of microsatellite instability in sporadic and inherited colorectal cancer, many studies have shown that microsatellite instability occurs in a wide variety of presumably sporadic tumors, including sporadic endometrial carcinomas. Like sporadic colorectal cancer, microsatellite instability has been found in approximately 20% of sporadic endometrioid endometrial cancers (BURKS et al. 1994; RISINGER et al. 1993). This observation is of special interest in light of the fact that endometrial cancer is the second most common tumor type occurring in HNPCC and that the endometrioid variant of endometrial cancer has other pathogenetic features in common with colorectal carcinoma.

To date, only limited studies have addressed the underlying genetic defect that generates instability of microsatellite sequences in sporadic cancers. One study thoroughly analyzed all four of the known human DNA mismatch repair genes in DNA isolated from transformed lymphoblastoid cell lines of ten patients with presumably sporadic MI-positive colorectal cancers (LIU et al. 1995). A germ line mutation was detected in only one case (mutation of *hMLH1*). The identified mutation was a nonsense mutation resulting in a severely truncated protein. Furthermore, no normal transcript was identified in the cell line generated from this patient's tumor. In retrospect, this patient, although not meeting the criteria for HNPCC, had two other family members with colorectal and endometrial cancer and the patient was less than 50 years old. Thus, it is possible that this patient is a member of an HNPCC family that had not been recognized previously as such. The investigators went on to study cell lines derived from MI-positive colorectal cancers arising in patients without significant family cancer histories. Only three of the seven cell lines harbored mutations in the four known DNA mismatch repair genes, all in *hMLH1*, and both copies of the gene were mutated somatically. A somewhat similar situation is seen in MI-positive sporadic endometrial carcinomas. Eight MI-positive primary endometrial carcinomas have been studied for the presence of mutations in all four known DNA mismatch repair genes (KATABUCHI et al., 1995). Only one tumor had a detectable somatic mutation and no germ line mutations were identified. The somatic mutation, in the *hMSH2* gene, is a splice site mutation that causes an exon to be skipped. The aberrant RNA processing creates a frameshift that results in a premature termination codon and, in turn, a truncated protein. Based on these results, it appears that germ line mutations in the known DNA mismatch repair genes in patients with presumably sporadic MI-positive colorectal and endometrial cancer are uncommon and that microsatellite instability in many sporadic forms of these cancers must be due to mutations in genes that have yet to be identified.

These results have significant ramifications for the use of MI assays in screening the general population for predisposition to colorectal cancer. When microsatellite instability was first identified in a subset of presumably sporadic colorectal cancers, it was thought that it might be useful in identifying potential members of HNPCC families. Since microsatellite instability assays are technically straightforward and fast, they appeared to offer a potentially cost-effective method for generalized screening. However, the recent data discussed above indicate that germ line mutations in the four known DNA mismatch repair genes in patients without a documented family cancer history are, at best, rare. Based on these results, it appears that generalized screening for microsatellite instability is not warranted. However, screening young patients with presumably sporadic colorectal cancers may be worthwhile, especially if the patient has a family member with colorectal cancer.

Finally, it is clear that additional genes exist that give rise to microsatellite instability. These genes may encode additional proteins involved in DNA mismatch repair, but it is likely that they may also encode proteins involved in replication, recombination, and other types of DNA repair processes. Obviously, the search for such genes is currently being vigorously pursued. In addition, if a mutator phenotype is required for the development of neoplasia then microsatellite instability must be only one manifestation of such a phenotype, as the vast majority of cancers lack instability of microsatellite DNA sequences. Thus, the nature of the mutator phenotype in the majority of tumors remains a mystery that will undoubtedly be a major focus of future investigation.

5 Changes in DNA Methylation Are Observed in Cancer Cells

DNA methylation involves covalent modification of DNA and as such can be considered a structural alteration of DNA in tumor cells. Changes in DNA methylation have been shown to influence several cellular processes that affect both DNA integrity and function, processes that go awry during tumorigenesis. Over the past several years, data have accumulated suggesting that changes in DNA methylation may play a role in tumor development and/or progression (reviewed in LAIRD and JAENISCH 1994). In mammals, DNA methylation occurs at the fifth carbon position of cytosine residues within CpG dinucleotides (ADAMS and LINDSAY 1993; GARDINER-GARDEN and FROMMER 1987). Although most of the CpG dinucleotides in the human genome are methylated, CpG dinucleotides are approximately fivefold underrepresented in the genome overall. Furthermore, CpG dinucleotides and 5-methylcytosine residues are unevenly distributed in the genome. Approximately 1%–2% of the human genome consists of "islands" of non-methylated CpG-rich stretches of DNA that are typically 1–2 kb in length located at the 5′ end of genes in association with the promoter. These non-methylated CpG islands are associated with all known housekeeping genes and with a number of genes that are expressed in a tissue-dependent fashion. Although housekeeping genes are ubiquitously expressed and thought not to be regulated by DNA methylation, several studies suggest an inverse correlation between DNA methylation and tissue-specific expression of certain genes, with the strongest correlation observed between gene activity and promoter hypomethylation (YEIVIN and RAZIN 1993). In general, methylated genes are inactive, while non-methylated genes are active. Thus DNA methylation provides a potential mechanism with which to inactivate specific genes (e.g., tumor suppressor genes) during tumorigenesis. Results of studies of cells in culture suggest that gene silencing by methylation can provide a selective advantage for cell growth (ANTEQUERA et al. 1990). Unfortunately, our understanding of the cellular processes regulating de novo methylation activity remains incomplete. Currently, only one DNA (cytosine-5) methyltransferase has been identified in human cells (YEN et al. 1992; BESTOR et al. 1988), and its major function appears to lie in maintenance of methylation status rather than de novo methylation.

Tumor cells exhibit both global changes in DNA methylation as well as alterations of methylation patterns in individual genes. Global hypomethylation has been reported in several tumor types including colorectal carcinoma (GOELZ et al. 1985), cervical carcinoma (KIM et al. 1994), seminomas (PELTOMAKI 1991), and others (RAO et al. 1989), and, in some cases, the degree of hypomethylation correlates with tumor progression (KIM et al. 1994; GAMA-SOSA et al. 1983). Hypomethylation of specific proto-oncogenes such as c-*MYC*, c-*FOS*, H-*RAS*, K-*RAS*, *ERB*-A1, and *BCL*-2 has also been observed, particularly in hepatocellular carcinomas and leukemias (reviewed in LAIRD and JAENISCH 1994).

Tumor cells also show regional increases in DNA methylation. For example, BAYLIN and his colleagues have found hypermethylation hot spots on chromosomes 3p, 11p, and 17p in several different types of human tumors (DE BUSTROS et al. 1988; MAKOS et al. 1992, 1993a, b). Moreover, other studies suggest that specific genes may be inactivated by hypermethylation in certain tumor types. For example, the *RB* tumor suppressor gene may be inactivated by hypermethylation in sporadic retinoblastomas (OHTANI-FUJITA et al. 1993; SAKAI et al. 1991), and, more recently, hypermethylation of regulatory sequences at the *GSTP*1 (glutathione S-transferase Pi-1) locus was detected in all human prostatic carcinomas studied but not in normal prostatic tissues or in tissues with benign prostatic hyperplasia (LEE et al. 1994). Expression of *GSTP*1 mRNA and protein was concomitantly reduced in the carcinomas. These findings may lead to several clinical applications. For example, since *GSTP*1 hypermethylation is the most common genetic alteration yet described in prostate cancer (in virtually 100% of cases), detection of this change (and/or loss of GSTP1 protein expression) may serve as an adjunctive tool in the diagnosis of prostate cancer. Also, the GSTP1 protein is an enzyme that is in part responsible for the detoxification of certain carcinogens. Expression of GSTP1 and related enzymes is in turn induced by certain foods such as cruciferous vegetables, suggesting a prostate cancer prevention strategy based on diets rich in foods that induce enzymes capable of compensating for loss of GSTP1 function.

In addition to the regional and global methylation changes noted in human cancers, it has long been recognized that CpG dinucleotides are hypermutable. That is, the estimated mutation rate of CpG dinucleotides is 10–40 times greater than that of other dinucleotides. The excess of mutations identified at CpG dinucleotides is attributable to transitions of CpG to TpG or CpA (SVED and BIRD 1990), many via spontaneous deamination of cytosine and especially 5-methylcytosine. Data are also accumulating suggesting that, in some cases, deamination may be facilitated by certain cellular enzymes, including a 5-methylcytosine-DNA glycosylase and/or the known DNA methyltransferase (reviewed in LAIRD and JAENISCH 1994).

6 Aneuploidy Frequently Characterizes Tumor Cells

Aneuploidy may be defined as an abnormal state in which the total DNA content of the aneuploid cells differs from the DNA content characteristic of that species, through the gain or loss of variable numbers of chromosomes or chromosome fragments. Normal somatic cells are diploid, that is, they contain a pair of each haploid ($n = 23$) chromosomes. In and of itself, diploidy does not imply normal chromosome structure or arrangement. Chromosomal losses may be balanced by chromosomal gains elsewhere in the genome. Similarly, aneuploidy is not observed only in malignant and pre-malignant cells. Certain forms of aneuploidy, including Klinefelter's syndrome, Turner's syndrome, and several trisomies, are compatible with near-normal growth and development. Aneuploidy may also arise from the gain of complete chromosomal sets. DNA indices (defined as the measurable ratio of abnormal to normal cellular DNA) up to 3.0 have been reported; however, the maximum DNA content that is compatible with cell survival remains unknown. The biological significance of aneuploidy has been recently reviewed (REW 1994).

Up to 75% of all tumors analyzed display aneuploidy. One feature of aneuploid tumors is that they typically contain only one or a few stable aneuploid populations of cells. In flow-cytometric analyses, the DNA profile of the aneuploid cells is typically superimposed on the DNA profile of diploid cells within the tumor. This diploid population usually represents a mix of non-malignant bystander cells (e.g., stromal cells, inflammatory cells, endothelial cells, etc.) and diploid tumor cells.

Several studies have assessed the value of aneuploidy as a marker of biological aggressiveness in various tumor types. Overall, conflicting views have emerged regarding the utility of ploidy data in predicting prognosis or directing clinical management. These different conclusions most likely reflect a combination of true biological differences between different series of tumors and technical differences in data collection and analysis. Thus far, the identification of aneuploidy in and of itself has proven to be of relatively little clinical utility. In general terms, while diploidy has been associated with a more favorable prognosis in some types of tumors (e.g., colorectal, breast, and prostatic carcinomas), many purely diploid tumors behave very aggressively.

It should be emphasized that diploidy and aneuploidy are not two distinct states, but part of a continuum from very subtle DNA changes to gross chromosomal abnormalities. In this context, the definition of aneuploidy may be somewhat arbitrary, since subtle but functionally important chromosomal alterations may escape detection with the laboratory techniques that are currently available. Given these caveats, it remains unclear whether there is a causal relationship between aneuploidy and cancer or whether aneuploidy simply reflects an epiphenomenon in genetically unstable tumor cells.

References

Aaltonen L, Peltomaki P, Leach F, Sistonen P, Pylkkanen L, Mecklin J, Jarvinen H, Powell S, Jen J, Hamilton S, Petersen G, Kinzler K, Vogelstein B, de la Chapelle A (1993) Clues to the pathogenesis of familial colorectal cancer. Science 260: 812–816

Adams RL, Lindsay H (1993) What is hemimethylated DNA? FEBS Lett 320: 243–245

Antequera F, Boyes J, Bird A (1990) High levels of de novo methylation and altered chromatin structure at CpG islands in cell lines. Cell 62: 503–514

Baker VV, Hatch KD, Shingleton HM (1988) Amplification of the c-myc proto-oncogene in cervical carcinoma. J Surg Oncol 39: 225–228

Barbacid M (1987) Ras genes. Annu Rev Biochem 56: 779–827

Bartram CR, De Klein A, Hagemeijer A, van Agthoven T, Geurts van Kessel A, Bootsma D, Grosveld G, Ferguson-Smith MA, Davies T, Stone M et al (1983) Translocation of c-abl oncogene correlates with the presence of a Philadelphia chromosome in chronic myelocytic leukaemia. Nature 306: 277–280

Bestor T, Laudano A, Mattaliano R, Ingram V (1988) Cloning and sequencing of a cDNA encoding DNA methyltransferase of mouse cells. The carboxyl-terminal domain of the mammalian enzymes is related to bacterial restriction methyltransferases. J Mol Biol 203: 971–983

Bishop JM (1985) Viral oncogenes. Cell 52: 301–354

Blackwood EM, Eisenman RN (1991) Max: a helix-loop-helix zipper protein that forms a sequence-specific DNA-binding complex with Myc. Science 251: 1211–1217

Boguski MS, McCormick F (1993) Proteins regulating Ras and its relatives. Nature 366: 643–654
Borden EC, Waalen J, Liberati AM, Grignani F (1993) Molecular diagnosis and monitoring of leukemia and lymphoma. Leuk Res 17: 1073–1078
Bos JL (1989) ras oncogenes in human cancer: a review. Cancer Res 49: 4682–4689
Bourne HR, Sanders DA, McCormick F (1990) The GTPase superfamily: a conserved switch for diverse cell functions. Nature 348: 125–132
Brash DE, Rudolph JA, Simon JA, Lin A, McKenna GJ, Baden HP, Halperin AJ, Ponten J (1991) A role for sunlight in skin cancer: UV-induced p53 mutations in squamous cell carcinoma. Proc Natl Acad Sci USA 88: 10124–10128
Brennan JA, Mao L, Hruban RH, Boyle JO, Eby YJ, Koch WM, Goodman SN, Sidransky D (1995) Molecular assessment of histopathological staging in squamous-cell carcinoma of the head and neck. N Engl J Med 332: 429–435
Brodeur GM, Seeger RC, Schwab M, Varmus HE, Bishop JM (1984) Amplification of N-myc in untreated human neuroblastomas correlates with advanced disease stage. Science 224: 1121–1124
Bronner CE, Baker SM, Morrison PT, Warren G, Smith LG, Lescoe MK, Kane M, Earabino C, Lipford J, Lindblom A, Tannergard P, Bollag RJ, Godwin AR, Ward DC, Nordenskjold M, Fishel R, Kolodner R, Liskay RM (1994) Mutation in the DNA mismatch repair gene homologue hMLH1 is associated with hereditary non-polyposis colon cancer. Nature 368: 258–260
Burks RT, Kessis TD, Cho KR, Hedrick L (1994) Microsatellite instability in endometrial carcinoma. Oncogene 9: 1163–1166
Caldas C, Hahn SA, Hruban RH, Redston MS, Yeo CJ, Kern SE (1994) Detection of K-ras mutations in the stool of patients with pancreatic adenocarcinoma and pancreatic ductal hyperplasia. Cancer Res 54: 3568–3573
Capon DJ, Chen EY, Levinson AD, Seeburg PH, Goeddel DV (1983) Complete nucleotide sequences of the T24 human bladder carcinoma oncogene and its normal homologue. Nature 302: 33–37
Cho YJ, Gorina S, Jeffrey PD, Pavletich NP (1994) Crystal structure of a p53 tumor suppressor DNA complex – understanding tumorigenic mutations. Science 265: 346–355
Cleary ML, Smith SD, Sklar J (1986) Cloning and structural analysis of cDNAs for bcl-2 and a hybrid bcl-2/immunoglobulin transcript resulting from the t(14;18) translocation. Cell 47: 19–28
Collins S, Groudine M (1982) Amplification of endogenous myc-related DNA sequences in a human myeloid leukaemia cell line. Nature 298: 679–681
Collins SJ, Kubonishi I, Miyoshi I, Groudine MT (1984) Altered transcription of the c-abl oncogene in K-562 and other chronic myelogenous leukemia cells. Science 225: 72–74
Cox EC (1976) Bacterial mutator genes and the control of spontaneous mutation. Annu Rev Genet 10: 135–156
Dalla-Favera R, Wong-Staal F, Gallo RC (1982) Onc gene amplification in promyelocytic leukaemia cell line HL-60 and primary leukaemic cells of the same patient. Nature 299: 61–63
de Bustros A, Nelkin BD, Silverman A, Ehrlich G, Poiesz B, Baylin SB (1988) The short arm of chromosome 11 is a "hot spot" for hypermethylation in human neoplasia. Proc Natl Acad Sci USA 85: 5693–5697
De Klein A, van Kessel AG, Grosveld G, Bartram CR, Hagmeijer A, Bootsma D, Spurr NK, Heisterkamp N, Groffen J, Stephenson JR (1982) A cellular oncogene is translocated to the Philadelphia chromosome in chronic myelocytic leukaemia. Nature 300: 765–767
De Klein A, Hagemeijer A, Bartram CR, Houwen R, Hoefsloot L, Carbonell F, Chan L, Barnett M, Greaves M, Kleihauer E et al (1986) bcr rearrangement and translocation of the c-abl oncogene in Philadelphia positive acute lymphoblastic leukemia. Blood 68: 1369–1375
DePinho RA, Schreiber-Agus N, Alt FW (1991) myc family oncogenes in the development of normal and neoplastic cells. Adv Cancer Res 57: 1–46
Der CJ, Krontiris TG, Cooper GM (1982) Transforming genes of human bladder and lung carcinoma cell lines are homologous to the ras genes of Harvey and Kirsten sarcoma viruses. Proc Natl Acad Sci USA 79: 3637–3640
Dougall WC, Qian X, Peterson NC, Miller MJ, Samanta A, Greene MI (1994) The neu-oncogene: signal transduction pathways, transformation mechanisms and evolving therapies. Oncogene 9: 2109–2123
Dowell SP, Hall PA (1994) The clinical relevance of the p53 tumour suppressor gene. Cytopathology 5: 133–145
Drebin JA, Link VC, Greene MI (1988) Monoclonal antibodies specific for the neu oncogene product directly mediate anti-tumor effects in vivo. Oncogene 2: 387–394
Dryja TP, Rapaport JM, Joyce JM, Petersen RA (1986) Molecular detection of deletions involving band q14 of chromosome 13 in retinoblastomas. Proc Natl Acad Sci USA 83: 7391–7394

Dunn JM, Phillips RA, Zhu X, Becker A, Gallie BL (1989) Mutations in the RB1 gene and their effects on transcription. Mol Cell Biol 9: 4596–4604

Fearon E, Vogelstein B (1990) A genetic model for colorectal tumorigenesis. Cell 61: 759–767

Fishel R, Lescoe MK, Rao MRS, Copeland NG, Jenkins NA, Garber J, Kane M, Kolodner R (1993) The human mutator gene homolog MSH2 and its association with hereditary nonpolyposis colon cancer. Cell 75: 1027–1038

Fishel R, Ewel A, Lee S, Lescoe MK, Griffith J (1994) Binding of mismatched microsatellite DNA sequences by the human msh2 protein. Science 266: 1403–1405

Francke U (1976) Retinoblastoma and chromosome 13. Cytogenet Cell Genet 16: 131–134

Friend S (1994) p53: a glimpse at the puppet behind the shadow play. Science 265: 334–335

Friend SH, Bernards R, Rogelj S, Weinberg RA, Rapaport JM, Albert DM, Dryja TP (1986) A human DNA segment with properties of the gene that predisposes to retinoblastoma and osteosarcoma. Nature 323: 643–646

Fuller GN, Bigner SH (1992) Amplified cellular oncogenes in neoplasms of the human central nervous system. Mutat Res 276: 299–306

Fung YK, Murphree AL, T'Ang A, Qian J, Hinrichs SH, Benedict WF (1987) Structural evidence for the authenticity of the human retinoblastoma gene. Science 236: 1657–1661

Gama-Sosa MA, Slagel VA, Trewyn RW, Oxenhandler R, Kuo KC, Gehrke CW, Ehrlich M (1983) The 5-methylcytosine content of DNA from human tumors. Nucleic Acids Res 11: 6883–6894

Gardiner-Garden M, Frommer M (1987) CpG islands in vertebrate genomes. J Mol Biol 196: 261–282

Garte SJ (1993) The c-myc oncogene in tumor progression. Crit Rev Oncog 4: 435–449

Goddard AD, Balakier H, Canton M, Dunn J, Squire J, Reyes E, Becker A, Phillips RA, Gallie BL (1988) Infrequent genomic rearrangement and normal expression of the putative RB1 gene in retinoblastoma tumors. Mol Cell Biol 8: 2082–2088

Goelz S, Vogelstein B, Hamilton S, Feinberg A (1985) Hypomethylation of DNA from benign and malignant human colon neoplasms. Science 228: 187–190

Greenblatt MS, Bennett WP, Hollstein M, Harris CC (1994) Mutations in the p53 tumor suppressor gene: clues to cancer etiology and molecular pathogenesis. Cancer Res 54: 4855–4878

Griesser H (1993) Applied molecular genetics in the diagnosis of malignant non-Hodgkin's lymphoma. Diagn Mol Pathol 2: 177–191

Groden J, Thliveris A, Samowitz W, Carlson M, Gelbert L, Albertsen H, Joslyn G, Stevens J, Spirio L, Robertson M et al (1991) Identification and characterization of the familial adenomatous polyposis coli gene. Cell 66: 589–600

Harris H (1988) The analysis of malignancy in cell fusion: the position in 1988. Cancer Res 48: 3302–3306

Heisterkamp N, Stephenson JR, Groffen J, Hansen PF, De Klein A, Bartram CR, Grosveld G (1983) Localization of the c-abl oncogene adjacent to a translocation break point in chronic myelocytic leukaemia. Nature 306: 239–242

Hockenbery DM, Oltvai ZN, Yin XM, Milliman CL, Korsmeyer SJ (1993) Bcl-2 functions in an antioxidant pathway to prevent apoptosis. Cell 75: 241–251

Ionov Y, Peinado MA, Malkhosyan S, Shibata D, Perucho M (1993) Ubiquitous somatic mutations in simple repeated sequences reveal a new mechanism for colonic carcinogenesis. Nature 363: 558–561

Jen J, Powell SM, Papadopoulos N, Smith KJ, Hamilton SR, Vogelstein B, Kinzler KW (1994) Molecular determinants of dysplasia in colorectal lesions. Cancer Res 54: 5523–5526

Kane DJ, Sarafian TA, Anton R, Hahn H, Gralla EB, Valentine JS, Ord T, Bredesen DE (1993) Bcl-2 inhibition of neural death: decreased generation of reactive oxygen species. Science 262: 1274–1277

Katabuchi H, Lambers A, Ronnett BR, Leach FS, Cho KR, Hedrick L (1995) Mutations in the known mismatch repair genes are not responsible for microsatellite instability in most sporadic endometrial carcinomas Cancer Res 55: 5556–5560

Khatib ZA, Matsushime H, Valentine M, Shapiro DN, Sherr CJ, Look AT (1993) Coamplification of the CDK4 gene with MDM2 and GLI in human sarcomas. Cancer Res 53: 5535–5541

Kim YI, Giuliano A, Hatch KD, Schneider A, Nour MA, Dallal GE, Selhub J, Mason JB (1994) Global DNA hypomethylation increases progressively in cervical dysplasia and carcinoma. Cancer 74: 893–899

Knudson A (1971) Mutation and cancer: statistical study of retinoblastoma. Proc Natl Acad Sci USA. 68: 820–823

Knudson A (1985) Hereditary cancer, oncogenes, and antioncogenes. Cancer Res 45: 1437–1443

Konopka JB, Watanabe SM, Witte ON (1984) An alteration of the human c-abl protein in K562 leukemia cells unmasks associated tyrosine kinase activity. Cell 37: 1035–1042

Krontiris TG, Cooper GM (1981) Transforming activity of human tumor DNAs. Proc Natl Acad Sci USA 78: 1181–1184

Laird PW, Jaenisch R (1994) DNA methylation and cancer. Hum Mol Genet 3: 1487–1495

Leach FS, Nicolaides NC, Papadopoulos N, Liu B, Jen J, Parsons R, Peltomaki P, Sistonen P, Aaltonen LA, Nystrom-Lahti M, Guan X-Y, Zhang J, Meltzer PS, Yu J-W, Kao F-T, Chen DJ, Cerosaletti KM, Fournier REK, Todd S, Lewis T, Leach RJ, Naylor SL, Weissenbach J, Mecklin J-P, Jarvinen H, Petersen GM, Hamilton SR, Green J, Jass J, Watson P, Lynch HT, Trent JM, de la Chapelle A, Kinzler K, Vogelstein B (1993) Mutations of a mutS homolog in hereditary nonpolyposis colorectal cancer. Cell 75: 1215–1226

Lee WH, Bookstein R, Hong F, Young LJ, Shew JY, Lee EY (1987) Human retinoblastoma susceptibility gene: cloning, identification, and sequence. Science 235: 1394–1399

Lee WH, Morton RA, Epstein JI, Brooks JD, Campbell PA, Bova GS, Isaacs WB, Nelson WG (1994) Cytidine methylation of regulatory sequences near the pi-class glutathione S-transferase gene accompanies human prostatic carcinogenesis. Proc Natl Acad Sci USA 91: 11733–11737

Lindblom A, Tannergard P, Werelius B, Nordenskjold M (1993) Genetic mapping of a second locus predisposing to hereditary non-polyposis colon cancer. Nature Genet 5: 279–282

Little CD, Nau MM, Carney DN, Gazdar AF, Minna JD (1983) Amplification and expression of the c-myc oncogene in human lung cancer cell lines. Nature 306: 194–196

Liu B, Nicolaides NC, Markowitz S, Willson JKV, Parsons RE, Jen J, Peltomaki P, Delachapelle A, Hamilton SR, Kinzler KW (1995) Mismatch repair gene defects in sporadic colorectal cancers with microsatellite instability. Nature Genet 9: 48–55

Loeb L (1991) Mutator phenotype may be required for multistage carcinogenesis. Cancer Res 51: 3075–3079

Lowy DR, Willumsen BM (1993) Function and regulation of ras. Annu Rev Biochem 62: 851–891

Lugo TG, Pendergast AM, Muller AJ, Witte ON (1990) Tyrosine kinase activity and transformation potency of bcr-abl oncogene products. Science 247: 1079–1082

Makos M, Nelkin BD, Lerman MI, Latif F, Zbar B, Baylin SB (1992) Distinct hypermethylation patterns occur at altered chromosome loci in human lung and colon cancer. Proc Natl Acad Sci USA 89: 1929–1933

Makos N, Nelkin BD, Chazin VR, Cavenee WK, Brodeur GM, Baylin SB (1993a) DNA hypermethylation is associated with 17p allelic loss in neural tumors. Cancer Res 53: 2715–2718

Makos M, Nelkin BD, Reiter RE, Gnarra JR, Brooks J, Isaacs W, Linehan M, and Baylin SB (1993b) Regional DNA hypermethylation at D17S5 precedes 17p structural changes in the progression of renal tumors. Cancer Res 53: 2719–2722

McCormick F (1993) Signal transduction. How receptors turn Ras on. Nature 363: 15–16

Modrich P (1991) Mechanisms and biological effects of mismatch repair. Annu Rev Genet 25: 229–253

Nicolaides N, Papadopoulos N, Liu B, Vogelstein B, Kinzler K (1994) Mutations in two PMS homologues in hereditary nonpolyposis colon cancer. Nature 371: 75–80

Nishisho I, Nakamura Y, Miyoshi Y, Miki Y, Ando H, Horii A, Koyama K, Utsunomiya J, Baba S, Hedge P, Markham A, Krush A, Petersen G, Hamilton S, Nilbert M, Levy D, Bryan T, Preisinger A, Smith K, Su L, Kinzler K, Vogelstein B (1991) Mutations of chromosome 5q21 genes in FAP and colorectal cancer patients. Science 253: 665–669

Nystromlahti M, Sistonen P, Mecklin JP, Pylkkanen L, Aaltonen LA, Weissenbach J, Delachapelle A, Peltomaki P (1994) Close linkage to chromosome 3p and conservation of ancestral founding haplotype in hereditary nonpolyposis colorectal cancer families. Proc Natl Acad Sci USA 91: 6054–6058

Ocadiz R, Sauceda R, Cruz M, Graef AM, Gariglio P (1987) High correlation between molecular alterations of the c-myc oncogene and carcinoma of the uterine cervix. Cancer Res 47: 4173–4177

Ohtani-Fujita N, Fujita T, Aoike A, Osifchin NE, Robbins PD, Sakai T (1993) CpG methylation inactivates the promoter activity of the human retinoblastoma tumor-suppressor gene. Oncogene 8: 1063–1067

Oshimura M, Kugoh H, Koi M, Shimizu M, Yamada H, Satoh H, Barrett JC (1990) Transfer of a normal human chromosome 11 suppresses tumorigenicity of some but not all tumor cell lines. J Cell Biochem 42: 135–142

Paik S (1992) Clinical significance of erbB-2 (HER-2/neu) protein. Cancer Invest 10: 575–579

Papadopoulos N, Nicolaides NC, Wei Y-F, Ruben SM, Carter KC, Rosen CA, Haseltine WA, Fleischmann RD, Fraser CM, Adams MD, Venter JC, Hamilton SR, Petersen GM, Watson P, Lynch HT, Peltomaki P, Mecklin J-P, de la Chapelle A, Kinzler KW, Vogelstein B (1994) Mutation of a mutL homolog in hereditary colon cancer. Science 263: 1625–1629

Parada LF, Tabin CJ, Shih C, Weinberg RA (1982) Human EJ bladder carcinoma oncogene is homologue of Harvey sarcoma virus ras gene. Nature 297: 474–478
Parsons R, Li GM, Longley MJ, Fang WH, Papadopoulos N, Jen J, Delachapelle A, Kinzler KW, Vogelstein B, Modrich P (1993) Hypermutability and mismatch repair deficiency in RER+ tumour cells. Cell 75: 1227–1236
Parsons R, Li G-M, Longley M, Modrich P, Liu B, Berk T, Hamilton SR, Kinzler KW, Vogelstein B (1995) Mismatch repair deficiency in phenotypically normal human cells. Science 268: 738–740
Peltomaki P (1991) DNA methylation changes in human testicular cancer. Biochim Biophys Acta 1096: 187–196
Peltomaki P, Aaltonen L, Sistonen P, Pylkkanen L, Mecklin J, Jarvinen H, Green J, Jass J, Weber J, Leach F, Petersen G, Hamilton S, de la Chapelle A, Vogelstein B (1993a) Genetic mapping of a locus predisposing to human colorectal cancer. Science 260: 810–812
Peltomaki P, Lothe RA, Aaltonen LA, Pylkkanen L, Nystrom-Lahti M, Seruca R, David L, Holm R, Ryberg D, Haugen A, Brogger A, Borresen AL, de la Chapelle A (1993b) Microsatellite instability is associated with tumors that characterize the hereditary non-polyposis colorectal carcinoma syndrome. Cancer Res 53: 5853–5855
Pendergast AM, Muller AJ, Havlik MH, Maru Y, Witte ON (1991) BCR sequences essential for transformation by the BCR-ABL oncogene bind to the ABL SH2 regulatory domain in a non-phosphotyrosine-dependent manner. Cell 66: 161–171
Perkins AS, Vande Woude GF (1993) Principles of molecular cell biology of cancer: oncogenes. In: Cancer: principles and practice of oncology. DeVita VT Jr, Hellman S, Rosenberg SA (eds) Lippincott, Philadelphia, pp 35–59
Perucho M, Goldfarb M, Shimizu K, Lama C, Fogh J, Wigler M (1981) Human-tumor-derived cell lines contain common and different transforming genes. Cell 27: 467–476
Potter MN, Cross NC, van Dongen JJ, Saglio G, Oakhill A, Bartram CR, Goldman JM (1993) Molecular evidence of minimal residual disease after treatment for leukaemia and lymphoma: an updated meeting report and review. Leukemia 7: 1302–1314
Powell SM, Zilz N, Beazerbarclay Y, Bryan TM, Hamilton SR, Thibodeau SN, Vogelstein B, Kinzler KW (1992) APC mutations occur early during colorectal tumorigenesis. Nature 359: 235–237
Powell SM, Petersen GM, Krush AJ, Booker S, Jen J, Giardiello FM, Hamilton SR, Vogelstein B, Kinzler KW (1993) Molecular diagnosis of familial adenomatous polyposis. N Engl J Med 329: 1982–1987
Press MF, Pike MC, Chazin VR, Hung G, Udove JA, Markowicz M, Danyluk J, Godolphin W, Sliwkowski M, Akita R et al (1993) Her-2/neu expression in node-negative breast cancer: direct tissue quantitation by computerized image analysis and association of overexpression with increased risk of recurrent disease. Cancer Res 53: 4960–4970
Pretlow TP, Brasitus TA, Fulton NC, Cheyer C, Kaplan EL (1993) K-ras mutations in putative preneoplastic lesions in human colon. J Natl Cancer Inst 85: 2004–2007
Rabbitts TH (1994) Chromosomal translocations in human cancer. Nature 372: 143–149
Rao PM, Antony A, Rajalakshmi S, Sarma DS (1989) Studies on hypomethylation of liver DNA during early stages of chemical carcinogenesis in rat liver. Carcinogenesis 10: 933–937
Rasheed BK, Bigner SH (1991) Genetic alterations in glioma and medulloblastoma. Cancer Metastasis Rev 10: 289–299
Reed JC (1994) Bcl-2 and the regulation of programmed cell death. J Cell Biol 124: 1–6
Rew DA (1994) Significance of aneuploidy. Br J Surg 81: 1416–1422
Risinger JI, Berchuck A, Kohler MF, Watson P, Lynch HT, Boyd J (1993) Genetic instability of microsatellites in endometrial carcinoma. Cancer Res 53: 1–4
Sakai T, Toguchida J, Ohtani N, Yandell DW, Rapaport JM, Dryja TP (1991) Allele-specific hypermethylation of the retinoblastoma tumor-suppressor gene. Am J Hum Genet 48: 880–888
Santoro M, Carlomagno F, Romano A, Bottaro DP, Dathan NA, Grieco M, Vecchio G, Matoskova B, Kraus MH, Difiore PP (1995) Activation of RET as a dominant transforming gene by germline mutations of MEN2A and MEN2B. Science 267: 381–383
Schwab M, Amler LC (1990) Amplification of cellular oncogenes: a predictor of clinical outcome in human cancer. Genes Chrom Cancer 1: 181–193
Sentman CL, Shutter JR, Hockenbery D, Kanagawa O, Korsmeyer SJ (1991) bcl-2 inhibits multiple forms of apoptosis but not negative selection in thymocytes. Cell 67: 879–888
Shibata D, Almoguera C, Forrester K, Dunitz J, Martin SE, Cosgrove MM, Perucho M, Arnheim N (1990) Detection of c-K-ras mutations in fine needle aspirates from human pancreatic adenocarcinomas. Cancer Res 50: 1279–1283

Shih C, Weinberg RA (1982) Isolation of a transforming sequence from a human bladder carcinoma cell line. Cell 29: 161–169

Shimizu M, Yokota J, Mori N, Shuin T, Shinoda M, Terada M, Oshimura M (1990) Introduction of normal chromosome 3p modulates the tumorigenicity of a human renal cell carcinoma cell line YCR. Oncogene 5: 185–194

Sidransky D, von Eschenbach A, Tsai YC, Jones P, Summerhayes I, Marshall F, Paul M, Green P, Hamilton SR, Frost P et al (1991) Identification of p53 gene mutations in bladder cancers and urine samples. Science 252: 706–709

Sidransky D, Tokino T, Hamilton SR, Kinzler KW, Levin B, Frost P, Vogelstein B (1992) Identification of ras oncogene mutations in the stool of patients with curable colorectal tumors. Science 256: 102–105

Slamon DJ, Godolphin W, Jones LA, Holt JA, Wong SG, Keith DE, Levin WJ, Stuart SG, Udove J, Ullrich A et al (1989) Studies of the HER-2/neu proto-oncogene in human breast and ovarian cancer. Science 244: 707–712

Slebos RJ, Kibbelaar RE, Dalesio O, Kooistra A, Stam J, Meijer CJ, Wagenaar SS, Vanderschueren RG, van Zandwijk N, Mooi WJ et al (1990) K-ras oncogene activation as a prognostic marker in adenocarcinoma of the lung. N Engl J Med 323: 561–565

Sparkes RS, Murphree AL, Lingua RW, Sparkes MC, Field LL, Funderburk SJ, Benedict WF (1983) Gene for hereditary retinoblastoma assigned to human chromosome 13 by linkage to esterase D. Science 219: 971–973

Strand M, Prolla TA, Liskay RM, Petes TD (1993) Destabilization of tracts of simple repetitive DNA in yeast by mutations affecting DNA mismatch repair. Nature 365: 274–276

Strasser A, Harris AW, Cory S (1991) bcl-2 transgene inhibits T cell death and perturbs thymic self-censorship. Cell 67: 889–899

Sved J, Bird A (1990) The expected equilibrium of the CpG dinucleotide in vertebrate genomes under a mutation model. Proc Natl Acad Sci USA 87: 4692–4696

Thibodeau SN, Bren G, Schaid D (1993) Microsatellite instability in cancer of the proximal colon. Science 260: 816–819

Trent JM, Stanbridge EJ, McBride HL, Meese EU, Casey G, Araujo DE, Witkowski CM, Nagle RB (1990) Tumorigenicity in human melanoma cell lines controlled by introduction of human chromosome 6. Science 247: 568–571

Tsujimoto Y, Cossman J, Jaffe E, Croce CM (1985) Involvement of the bcl-2 gene in human follicular lymphoma. Science 228: 1440–1443

Urban T, Ricci S, Grange JD, Lacave R, Boudghene F, Breittmayer F, Languille O, Roland J, Bernaudin JF (1993) Detection of c-Ki-ras mutation by PCR/RFLP analysis and diagnosis of pancreatic adenocarcinomas. J Natl Cancer Inst 85: 2008–2012

Vaux DL, Cory S, Adams JM (1988) Bcl-2 gene promotes haemopoietic cell survival and cooperates with c-myc to immortalize pre-B cells. Nature 335: 440–442

Vogelstein B, Kinzler KW (1992) p53 function and dysfunction. Cell 70: 523–526

Vogelstein B, Fearon E, Hamilton S, Kern S, Preisinger A, Leppert M, Nakamura Y, White R, Smits A, Bos J (1988) Genetic alterations during colorectal-tumor development. N Engl J Med 319: 525–532

Weinberg RA (1994) Oncogenes and tumor suppressor genes. Ca Cancer J Clin 44: 160–170

Weissman BE, Saxon PJ, Pasquale SR, Jones GR, Geiser AG, Stanbridge EJ (1987) Introduction of a normal human chromosome 11 into a Wilms' tumor cell line controls its tumorigenic expression. Science 236: 175–180

Yandell DW, Campbell TA, Dayton SH, Petersen R, Walton D, Little JB, McConkie-Rosell A, Buckley EG, Dryja TP (1989) Oncogenic point mutations in the human retinoblastoma gene: their application to genetic counseling. N Engl J Med 321: 1689–1695

Yeivin A, Razin A (1993) DNA methylation: molecular biology and biological significance Birkhauser, Basel

Yen RW, Vertino PM, Nelkin BD, Yu JJ, el-Deiry W, Cumaraswamy A, Lennon GG, Trask BJ, Celano P, Baylin SB (1992) Isolation and characterization of the cDNA encoding human DNA methyltransferase. Nucleic Acids Res 20: 2287–2291

Yunis JJ, Tanzer J (1993) Molecular mechanisms of hematologic malignancies. Crit Rev Oncog 4: 161–190

Zhou DJ, Gonzalez-Cadavid N, Ahuja H, Battifora H, Moore GE, Cline MJ (1988) A unique pattern of proto-oncogene abnormalities in ovarian adenocarcinomas. Cancer 62: 1573–1576

Subject Index

Current Topics in Microbiology and Immunology

Volumes published since 1989 (and still available)

Vol. 181: **Russell, Stephen W.; Gordon, Siamon (Eds.):** Macrophage Biology and Activation. 1992. 42 figs. IX, 299 pp. ISBN 3-540-55293-6

Vol. 182: **Potter, Michael; Melchers, Fritz (Eds.):** Mechanisms in B-Cell Neoplasia. 1992. 188 figs. XX, 499 pp. ISBN 3-540-55658-3

Vol. 183: **Dimmock, Nigel J.:** Neutralization of Animal Viruses. 1993. 10 figs. VII, 149 pp. ISBN 3-540-56030-0

Vol. 184: **Dunon, Dominique; Mackay, Charles R.; Imhof, Beat A. (Eds.):** Adhesion in Leukocyte Homing and Differentiation. 1993. 37 figs. IX, 260 pp. ISBN 3-540-56756-9

Vol. 185: **Ramig, Robert F. (Ed.):** Rotaviruses. 1994. 37 figs. X, 380 pp. ISBN 3-540-56761-5

Vol. 186: **zur Hausen, Harald (Ed.):** Human Pathogenic Papillomaviruses. 1994. 37 figs. XIII, 274 pp. ISBN 3-540-57193-0

Vol. 187: **Rupprecht, Charles E.; Dietzschold, Bernhard; Koprowski, Hilary (Eds.):** Lyssaviruses. 1994. 50 figs. IX, 352 pp. ISBN 3-540-57194-9

Vol. 188: **Letvin, Norman L.; Desrosiers, Ronald C. (Eds.):** Simian Immunodeficiency Virus. 1994. 37 figs. X, 240 pp. ISBN 3-540-57274-0

Vol. 189: **Oldstone, Michael B. A. (Ed.):** Cytotoxic T-Lymphocytes in Human Viral and Malaria Infections. 1994. 37 figs. IX, 210 pp. ISBN 3-540-57259-7

Vol. 190: **Koprowski, Hilary; Lipkin, W. Ian (Eds.):** Borna Disease. 1995. 33 figs. IX, 134 pp. ISBN 3-540-57388-7

Vol. 191: **ter Meulen, Volker; Billeter, Martin A. (Eds.):** Measles Virus. 1995. 23 figs. IX, 196 pp. ISBN 3-540-57389-5

Vol. 192: **Dangl, Jeffrey L. (Ed.):** Bacterial Pathogenesis of Plants and Animals. 1994. 41 figs. IX, 343 pp. ISBN 3-540-57391-7

Vol. 193: **Chen, Irvin S. Y.; Koprowski, Hilary; Srinivasan, Alagarsamy; Vogt, Peter K. (Eds.):** Transacting Functions of Human Retroviruses. 1995. 49 figs. IX, 240 pp. ISBN 3-540-57901-X

Vol. 194: **Potter, Michael; Melchers, Fritz (Eds.):** Mechanisms in B-cell Neoplasia. 1995. 152 figs. XXV, 458 pp. ISBN 3-540-58447-1

Vol. 195: **Montecucco, Cesare (Ed.):** Clostridial Neurotoxins. 1995. 28 figs. XI., 278 pp. ISBN 3-540-58452-8

Vol. 196: **Koprowski, Hilary; Maeda, Hiroshi (Eds.):** The Role of Nitric Oxide in Physiology and Pathophysiology. 1995. 21 figs. IX, 90 pp. ISBN 3-540-58214-2

Vol. 197: **Meyer, Peter (Ed.):** Gene Silencing in Higher Plants and Related Phenomena in Other Eukaryotes. 1995. 17 figs. IX, 232 pp. ISBN 3-540-58236-3

Vol. 198: **Griffiths, Gillian M.; Tschopp, Jürg (Eds.):** Pathways for Cytolysis. 1995. 45 figs. IX, 224 pp. ISBN 3-540-58725-X

Vol. 199/I: **Doerfler, Walter; Böhm, Petra (Eds.):** The Molecular Repertoire of Adenoviruses I. 1995. 51 figs. XIII, 280 pp. ISBN 3-540-58828-0

Vol. 199/II: **Doerfler, Walter; Böhm, Petra (Eds.):** The Molecular Repertoire of Adenoviruses II. 1995. 36 figs. XIII, 278 pp. ISBN 3-540-58829-9

Vol. 199/III: **Doerfler, Walter; Böhm, Petra (Eds.):** The Molecular Repertoire of Adenoviruses III. 1995. 51 figs. XIII, 310 pp. ISBN 3-540-58987-2

Vol. 200: **Kroemer, Guido; Martinez-A., Carlos (Eds.):** Apoptosis in Immunology. 1995. 14 figs. XI, 242 pp. ISBN 3-540-58756-X

Vol. 201: **Kosco-Vilbois, Marie H. (Ed.):** An Antigen Depository of the Immune System: Follicular Dendritic Cells. 1995. 39 figs. IX, 209 pp. ISBN 3-540-59013-7

Vol. 202: **Oldstone, Michael B. A.; Vitković, Ljubiša (Eds.):** HIV and Dementia. 1995. 40 figs. XIII, 279 pp. ISBN 3-540-59117-6

Vol. 203: **Sarnow, Peter (Ed.):** Cap-Independent Translation. 1995. 31 figs. XI, 183 pp. ISBN 3-540-59121-4

Vol. 204: **Saedler, Heinz; Gierl, Alfons (Eds.):** Transposable Elements. 1995. 42 figs. IX, 234 pp. ISBN 3-540-59342-X

Vol. 205: **Littman, Dan R. (Ed.):** The CD4 Molecule. 1995. 29 figs. XIII, 182 pp. ISBN 3-540-59344-6

Vol. 206: **Chisari, Francis V.; Oldstone, Michael B. A. (Eds.):** Transgenic Models of Human Viral and Immunological Disease. 1995. 53 figs. XI, 345 pp. ISBN 3-540-59341-1

Vol. 207: **Prusiner, Stanley B. (Ed.):** Prions Prions Prions. 1995. 42 figs. VII, 163 pp. ISBN 3-540-59343-8

Vol. 208: **Farnham, Peggy J. (Ed.):** Transcriptional Control of Cell Growth. 1995. 17 figs. IX, 141 pp. ISBN 3-540-60113-9

Vol. 209: **Miller, Virginia L. (Ed.):** Bacterial Invasiveness. 1996. 16 figs. IX, 115 pp. ISBN 3-540-60065-5

Vol. 210: **Potter, Michael; Rose, Noel R. (Eds.):** Immunology of Silicones. 1996. 136 figs. XX, 430 pp. ISBN 3-540-60272-0

Vol. 211: **Wolff, Linda; Perkins, Archibald S. (Eds.):** Molecular Aspects of Myeloid Stem Cell Development. 1996. 98 figs. XIV, 298 pp. ISBN 3-540-60414-6

Vol. 212: **Vainio, Olli; Imhof, Beat A. (Eds.):** Immunology and Developmental Biology of the Chicken. 1996. 43 figs. IX, 281 pp. ISBN 3-540-60585-1

Vol. 213/I: **Günthert, Ursula; Birchmeier, Walter (Eds.):** Attempts to Understand Metastasis Formation I. 1996. 35 figs. XV, 293 pp. ISBN 3-540-60680-7

Vol. 213/II: **Günthert, Ursula; Birchmeier, Walter (Eds.):** Attempts to Understand Metastasis Formation II. 1996. 33 figs. XV, 288 pp. ISBN 3-540-60681-5

Vol. 213/III: **Günthert, Ursula; Schlag, Peter M.; Birchmeier, Walter (Eds.):** Attempts to Understand Metastasis Formation III. 1996. 14 figs. XV, 262 pp. ISBN 3-540-60682-3

Vol. 214: **Kräusslich, Hans-Georg (Ed.):** Morphogenesis and Maturation of Retroviruses. 1996. 34 figs. XI, 344 pp. ISBN 3-540-60928-8

Vol. 215: **Shinnick, Thomas M. (Ed.):** Tuberculosis. 1996. 46 figs. XI, 307 pp. ISBN 3-540-60985-7

Vol. 216: **Rietschel, Ernst Th.; Wagner, Hermann (Eds.):** Pathology of Septic Shock. 1996. 34 figs. X, 321 pp. ISBN 3-540-61026-X

Vol. 217: **Jessberger, Rolf; Lieber, Michael R. (Eds.):** Molecular Analysis of DNA Rearrangements in the Immune System. 1996. 43 figs. IX, 224 pp. ISBN 3-540-61037-5

Vol. 218: **Berns, Kenneth I.; Giraud, Catherine (Eds.):** Adeno-Associated Virus (AAV) Vectors in Gene Therapy. 1996. 38 figs. IX,173 pp. ISBN 3-540-61076-6

Vol. 219: **Gross, Uwe (Ed.):** Toxoplasma gondii. 1996. 31 figs. XI, 274 pp. ISBN 3-540-61300-5

Vol. 220: **Vogt, Peter K.; Rauscher, Frank J. III (Eds.):** Chromosomal Translocations and Oncogenic Transcription Factors. 1997. 26 figs. X, 180 pp. ISBN 3-540-61402-8

Springer and the environment

At Springer we firmly believe that an international science publisher has a special obligation to the environment, and our corporate policies consistently reflect this conviction.

We also expect our business partners – paper mills, printers, packaging manufacturers, etc. – to commit themselves to using materials and production processes that do not harm the environment. The paper in this book is made from low- or no-chlorine pulp and is acid free, in conformance with international standards for paper permanency.

Springer

Printing: Saladruck, Berlin
Binding: Buchbinderei Lüderitz & Bauer, Berlin